FORESTRY RESEARCH EXTENSION

CHALLENGES & STRATEGIES

OTHER RELATED BOOKS

Biodiversity of Ranthambhore Tiger Reserve	*V. Singh*
Multipurpose Plants	*P.K. Srivastava*
Multipurpose trees in the Tropics: Management and Imporvement Strategies	*V.P. Tewari*
Neem in Sustainable Agriculture	*Narwal, S.S.*
Nursery & Plantation Practices in Forestry (2nd Ed.)	*V. Kumar*
Pests of Forest Importance and their Management	*B.K. Tyagi*
Pioneer Research in Extension Education	*A. Singh*
Plant Reproduction	*T. Pullaiah*
Plant Tissue Culture	*P. Dwivedi*
Plant Tissue Culture - Theory & Practicals	*T. Pullaiah*
Plant Tissue Culture - Theory & Techniques	*S. Kumar*
Practical Manual on Plantation Forestry	*Pankaj Panwar*
Tropical Forest Ecosystem: Structure & Function	*V.P. Singh*

For Details visit : www.scientificpub.com

FORESTRY RESEARCH EXTENSION

CHALLENGES & STRATEGIES

C.S. Dange
Seema Kumar
Arid Forest Research Institute, Jodhpur (Raj.)

SCIENTIFIC
PUBLISHERS (INDIA)

Published by:

Scientific Publishers (India)
5 A, New Pali Road, P.O. Box 91
Jodhpur 342 001 (India)
E-mail: info@scientificpub.com
Website: www.scientificpub.com

Scientific Publishers (India)
4806/24, Ansari Road, Daryaganj
New Delhi - 110 002 (India)

Print: 2013

ISBN: 978-81-7233-596-0

Lasertype set : Rajesh Ojha
Printed in India

FOREWORD

The Indian Council of Forestry Research and Education (ICFRE) has mission to generate, preserve, disseminate and promote knowledge, technologies and solutions for addressing issues related to Forests, Environment and wild life and to facilitate management of these resources on a sustained basis for benefit of the human kind through Research, Education and Extension. It endeavors to bridge the gap between basic and applied research through its Research Institutes and Centres. The Arid Forest Research Institute (AFRI), Jodhpur and other Research Institutes/Centres under ICFRE have developed many appropriate technologies to enhance productivity of our degraded lands both within and outside notified forest areas through efficient land use. Some of these have already been adopted by user groups including Forest Departments and Farmers. However, appropriate extension methodologies for effectively transferring the developed technologies to the end users are still in the developing stage and need refinement. Hence, the users are deprived of the full benefits of these technologies and, in turn, are not able to derive maximum benefits for improving productivity of their lands and enhancing their economic status. Keeping the XI Five Year Plan perspective priorities in view, we intend to sensitize stakeholders particularly farmers and villagers through demonstration of developed technologies on their farms in selected villages. Establishment of at least two Van Vigyan Kendras (VVKs) in each State and one in each Union Territory to showcase and demonstrate these technologies in and around these VVKs is being attempted. The relevant literature would be made available in regional languages.

The harsh climatic conditions, erratic rainfall and frequent draughts in the desert areas have created acute shortage of fuel wood, fodder, timber and other minor forest products making the life of local inhabitant very cumbersome. This exerts immense pressure on the existing sparse vegetational diversity of the area. There is a huge gap between demand and supply of these native resources.

To bridge this gap and to identify these issues, AFRI organized a regional workshop on "Challenges in Forestry Research Extension" during 18-19th October, 2005 and brought together the scientists, subject-matter specialists, Forest Managers and all the user groups to deliberate upon these issues and suggest ways and means for effective and workable methodologies for Extension of these technologies.

Future forestry research technology dissemination strategy necessitates applying institutional integrated approach in a coordinated manner through effective nodal extension wing having well versed and experienced technical staff. In order to achieve this objective, involvement of concerned forestry research personnel interfaced with development of technology and its extension under Lab to Land Programme is a must. This can be realized only when each experiment/project has provision of extension component as an integral part.

The book has embarked on wide range of topics related to forestry research extension presented by various subject-matter specialists during the different sessions of the workshop. I express my happiness and satisfaction on successful organization of the workshop and appreciate the efforts of all persons involved in its organization, particularly the presenters and all the members of the organizing committee who worked tirelessly to make it a success.

I congratulate the organizers and the editors for their sincere efforts to bring out this book as an outcome of the workshop and I hope that it will become a useful tool in furthering the cause of forestry research extension effectively, linking all stakeholders with forestry research development in one way or the other.

Dr. R. L. Srivastava I.F.S.
DIRECTOR
A.F.R.I., Jodhpur

PREFACE

Extension of forestry research development has remained under utilized for want of effective dissemination linkages with the user groups to the required extent. It has not been able to witness much focused progress despite rapid advancement in the information technology. Systematic concerted effort has to be made on the status of comprehensive extension on a regular sustainable basis through development of suitable site-specific demonstrative models with due participation of all stake-holders. Different extension methodologies and strategies shall have to be evolved to address different categories/strata of the target groups. The extension programme shall have to be supported with adequate funding, technical staff, machinery/equipments and infrastructure for expected extension of developed technologies. Effective extension wing coupled with effective planning, streamlined and standardized procedures can deliver essential results, growth and development.

One has to gain professionalism i.e., scientific and technical competence, bring out discernible, reliable, reproducible and demand driven technological packages in professional temperament. Simple techniques through multi-disciplinary, cohesive and integrated consortium mode shall have to be evolved with sincerity, integrity, dedication and honest working for their dissemination keeping pace with the changing times by having frequent access to the user.

The regional workshop on "Challenges in Forestry Research Extension" at AFRI Jodhpur from 18-19th October, 2005 was one such approach in bringing together the subject matter specialists and other participants on one platform so as to give them an opportunity to interact and devise recommendations for a framework of extension research, a useful tool to address the various multi-tier categories/strata of the target groups. It is our belief that this book shall prove to be a backdrop for immediate immense efficacy and result in site specific extension on regular basis.

It had become possible as a result of the valuable advisory support received from ICFRE and the willing participation of regional research extension professionals although its logical follow-up has to go a long way to make forestry research extension a grand success. We are thankful to all AFRI personnel in general and particularly those who could make all out efforts in smooth conduct of the workshop. Especially, We convey our gratitude to all the steadfast and eminent representative participants from I.C.F.R.E., Dehradun; C.A.Z.R.I., Jodhpur; Rajasthan Ayurved University, Jodhpur; State Remote Sensing Application Centre, Jodhpur; Desert Medicinal Research Centre, Jodhpur; Gujarat, Rajasthan and Dadra & Nagar Haveli Forest Departments; Rajasthan Agricultural University, Bikaner; Maharana Pratap University of Agriculture & Technology, Udaipur; N.G.O.s, Progressive Farmers and Medi-culturists for sparing their valuable time and enriching the research, scientific and technical forestry community with their knowledge, experience and vision for future dissemination by way of presentations and discussions held during the said workshop. They all did it with a sense of duty, devotion, commitment and conviction. We hope that success lies in our effective dissemination and shall be more than expected in time to come.

C.S. Dange, I.F.S.
Smt. Seema Kumar

CONTENTS

A CONCEPT PAPER ON FORESTRY EXTENSION

Mehar Singh
ADG (Media and Publication Division)
Directorate of Extension, I.C.F.R.E., Dehradun

ABSTRACT

Forestry research extension in India revolves around silvicultural conferences, quini-quinnial programmes and periodical liaison meetings. To begin with silvicultural conferences were held after 5 years. Silvicultural conferences were used to organize for discussion and dissemination of forestry researches. However, forestry extension remained confined to state forest departments until social forestry programme started in early 1970s. In 1994-95 an extension directorate was established in Indian Council of Forestry Research & Education under World Bank funded forestry research, education and extension project (FREEP). However, extension in the forestry sector had a back seat and more attention was given to forestry research in the recent past. ICFRE established extension departments at the level of ICFRE and also in its institutes. Regional workshops in different institutes of Indian Council of Forestry Research & Education such as this are being organized in order to understand the present condition of extension programmes in ICFRE and in the country as a whole in the forestry sector.

Thus, forestry research extension is a process used for communication of relevant and useful information to the stakeholders. It involves imparting of knowledge skills and attitudes for proper utilization of information in such a way that life quality of the concerned stakeholder is improved. Essentially, the purpose of the extension is to help the stakeholders in decision making skills so that they can organize, operate and participate in utilization of scarce resources for maximum benefit. This manuscript provides historical background of forestry research extension in the Indian Council of Forestry Research and Education.

DEFINITION

Forestry extension or for that matter any extension involves two elements. Firstly, it is a process used for communication of relevant and useful information to the stakeholders. Secondly, it involves imparting of knowledge, skills and attitudes for proper utilization of information in such a way that life quality of the concerned stakeholder is improved. Essentially, the purpose of the extension is to help the stake holders in decision making skills so that they can organize, operate and participate in utilization of scarce resources for maximum benefit.

History

The root of extension extends to the Renaissance period when there was an attempt to link the needs of the human life to the process of education. Modern science in 16th and 17th Century revealed the principles of working of nature. In the beginning astronomy and mechanics were the main branches of knowledge, which subsequently were followed by physics, chemistry and biology. Various laws governing the motion of matter at different levels in the above mentioned branches of knowledge were utilized in improving the life of society. It was felt that knowledge of nature and various inventions and discoveries must be translated into attainment of better quality of life.

Development in the field of forestry is closely associated with those in the field of agriculture. Because of these linkages one of the objectives of the first Forest policy of India (1894) envisages that forestry should be managed in order to improve the agriculture productivity of the country. This objective continued in the second forest policy of India (1952).

In the beginning, more attention was paid to agricultural development in order to meet the challenge of burgeoning population of the country. Therefore, it is natural that first attempts in an extension activity were in the field of agriculture.

Earliest agricultural schools were opened in 1779 in Hungary at Jarwas and then at Nagy – Mychlos in 1786 at Kozthely in 1797 (True, 1928 tree law page 3). Agricultural society was the pioneer in agricultural extension both in Europe and America (True, 1928). In 1973 first society was set up in Scotland and then in 1744 American Philosophy Society produced public articles on agricultural subjects. The committee on agriculture of New York Assembly in 1843 suggested that "the legislature might authorize the state agricultural society to employ a practical and scientific farmer to give public lectures through out the state upon practical and scientific knowledge (True, 1928, P-24)." The use of the proper term "Extension" can be traced back to 1866 in England which has its origin with the university extension system, first in Cambridge and then in Oxford university. The term extension education was first used by Cambridge University in 1873 to describe educational innovations.

Inception of agricultural extension education in third world countries is a post Second World War phenomenon. In most of these countries, extension programmes started with external assistance, mainly from U.S.A. In these countries, the extension education programmes were mainly a top down approach, unlike bottom up approach in case of European and North American countries. This characteristic has been found to be a major cause of failure of extension programs in Third World countries. They had a severe shortage of trained personnel and the personnel looking after extension activities were found more involved in administrative and regulatory activities. Clifford Wharton in 1983, President of State University of New York and former chairman of the Board of International Food & Agriculture Development admitted that the most unsuccessful efforts have been in the achievement of delivery of scientific and technical knowledge to the million farm producers of the third world countries.

Different approaches have been adopted over a period to address the problem of agricultural extension. Firstly, the conventional approach adopted the discharging of extension function by the Ministry of Agriculture. Consequently, the administrative functions rather than technical predominated.

To improve upon this problem, World Bank introduced an agricultural extension system which will have focus on training and visits. This system came to be called as T&V system.

Other alternative approach is university extension system. In this system the extension is organized under higher educational institutions. This kind of system is prevalent in North American countries.

Under conventional agricultural extension approach, the main objective was increasing agricultural production. As the above mentioned activity implies, the object of this system was to increase the production and there was no emphasis on increasing the incomes of farmers. It was assumed as the production will increase the lot of farmers will also improve. However, it was found that even though agricultural productivity was increased, the condition of the farmers continued to deteriorate. Main elements of extension *i.e.* communication and education were lacking in this approach. The extension personnel worked more as low level administrators than educators.

The training and visit system incorporates increase in farmer's production and income as one of its objectives. In addition to other objectives such as to improve the organization and extension through a single and direct line of technical and administrative control, to provide intensive coverage by reducing the number of families to be covered by the extension worker, to improve technical skills and knowledge of extension worker through regular training, by defining the contents of job of extension worker clearly and above all to avoid education and extension services by integrating the extension in a single line of control.

This system fixes the extension worker: farmer ratio as 1:800 on an average and thereby makes extension activity more accountable and specific. This is further ensured by prescribing fortnightly visits of extension worker on a fixed schedule known to farmers, supervisors and technical staff. Focus of extension in this system, to begin with, is to transfer low cost and low risk technology to farmers.

The main criticism of this system is that it is too much top down oriented and does not allow participation of the farmer, too rigid in terms of schedule, too much labour intensive, and does not make use of mass media methods of communication etc.

The University extension system mostly practiced in the United States is with an objective to improve the efficiency of the agriculture production, farmer's income and rural welfare and also prescribes working of communities for improving the community's living conditions.

In this case extension system is organized at state level at one or more of the state land grant university. This system has been partly followed in India and Philippines also. The project in this case is educational with an orientation to using research to help people identify and solve problems.

Community development and production system is another extension system with an objective of community and market development. Generally such commodities are produced for export such as tea, rubber and sugar. However, it can also be for domestic production e.g., milk. In this case quality is the most important factor. The producer has not much choice but to follow to use the inputs and follow the technical recommendations provided to him. The role of extension personnel in such cases is of providing recommended technology, message and other agro-services.

Forestry research extension in India revolves around silvicultural conferences, Quini-quinnial programmes and periodical liaison meetings. To begin with silvicultural conferences were held after 5 years. Quini-quinnial programmes were just like sectorial 5 year plans and liaison meetings were held to review the working between Forest Research Institute and State Forest Departments. Eleven silvicultural conferences were organized between 1818 and 1916. Forestry extension remained confined to State Forest Departments until social forestry programme started in early 1970s. In 1994-95 an extension directorate was established in Indian Council of Forestry Research & Education under World Bank funded forestry research, education and extension project (FREEP). However, extension in the forestry sector had a back seat and more attention was given to forestry research in the recent past. ICFRE established extension departments at the level of ICFRE and also in its institutes. However, because of non-availability of resources, financial as well as human, much is desired to be done in the field of forestry extension.

Present Condition

Regional workshops in different institutes of Indian Council of Forestry Research & Education such as this are being organized in order to understand the present condition of extension programmes in ICFRE and in the country as a whole in the forestry sector. It is a hope that the present regional workshop will come out with the understanding of present situation of extension education and recommendat-ions to achieve the desired goal which can be discussed thoroughly at the national level on completion of these regional workshops.

STRATEGIES FOR ADAPTATION OF FORESTRY- AGROFORESTRY RESEARCH FOR THE ENDUSERS

L.N. Harsh
Principal Scientist, CAZRI, Jodhpur-342 003

ABSTRACT

During mid sixties India was importing around 5-10 million tons of food annually. But due to cutting edge of science, strong political will coupled with appropriate policy interventions and hard labour of Indian farmers, the country has achieved four-fold increase (from 50 to 200 mt.) in food grain against the three fold increase in population (from 330-1000 million). However, incase of forestry, the achievement was abysmally low, though in-spite of good efforts and from time to time framing of new policies, the participation of people is minimum. There may be many causes for low acceptance of forestry/agroforestry.

Causes

The long gestation period for economic returns from the trees. People mentality is that plantation of trees are the job of forest department.

Complexity of innovation

- Effect on farmer's flexibility.
- Implementation cost: Capital out lay & Intellectual Duflay.

- Risk and uncertainty etc.

Keeping in view of above problems if strategies developed which can instigate the people to accept farm forestry/agroforestry.

Strategies

- Collaboration between researchers, Extension and landholders.
- Identifying economic and other benefits.
- Diagnosis and Design.
- The Extension and Advisory Spectrum.
    - Information access and technology transfer through Television, Radio and rural press, Information sheets and booklets.
    - One to one advise or information exchange.
    - Formal and structured education and training (Field days, Agroforestry Expo., University Courses, Short courses such as Master Tree Grovers Program etc.).

Adopting of above strategies, CAZRI, Jodhpur has adopted 10 villages in five districts of arid region. The results of the five years of the study has shown very encouraging success and economy of the people have increased from empty bowl to self-reliance.

Introduction

During mid sixties India was importing around 5-10 million tonnes of food annually. But due to cutting edge of science, strong political will coupled with appropriate policy intervention and hard labour of Indian farmers the country has achieved four fold increase (50 to 200 mt) in food grain against the three fold increase in population (from 300 to 1000 million). However, in case of forestry sector, inspite of concentrated efforts by the government, the increase in forest area is abysmally low (19.29%) (FSI, 2003). But in the recent past, due to commercialization of forestry sector, about 6% of forest cover is reported outside the forest areas (Chandra, 2003), as against people's participation in agriculture in forestry sector it is very low. The poor participation in forestry sector is due to many reasons. Black *et al.* (2000) studied the causes for poor participation of people in tree plantation. These are

- **Long Gestation Period**
    - From most of the trees, returns are after 7-10 years as compared to crops (3-5 months). Poor farmers can't wait for long period.
- **Tree plantation**
    - In general, people considered that plantation of tree is the job of forest department and they only know the utilization. Therefore, trees are not planted by the farmers.
- **Complexity of Innovation**
    - Silvicultural / Agroforestry is a complex subject in which farmers must know the effect of tree on the companion crops.
    - Lack of integration of a diverse range of skills, management and knowledge:
        - Silvicultural knowledge and skills.
        - Timber harvesting, marketing and processing.
        - Tree – Crop interactions
- **Effect on farmers flexibility**
    - Trees take much longer to reach maturity (10-15 yrs).
    - Due to changes in the market conditions, farm can't change the tree species in short period as in

case of agriculture crops. In agriculture, if a farmer sees that wheat rate has gone down, than in the next season, he replaces the crop but in trees he cannot uproot the trees. Therefore, he has to sell his produce at a lower cost. For example, poplar used to be sold @ Rs. 1200/t at one time but recently its price has come down to Rs. 400-600 / t.

- **Implementation cost**
  - **Capital out lay:** High establishment costs coupled with a long wait for return on the investment, pose a major problem for many farmers, specially if they must borrow to finance the venture.
  - **Intellectual out lay:** It means new learning by the farmers. Old generation unwilling or unable to learn new competencies. Most of the farmers have in the mind that growing of trees in agricultural farm will take away all the benefits of the crops.
- **Risk and Uncertainty**

Various risks and uncertainties associated with Farm Forestry / Agroforestry

  - Weather
  - Tree pests and diseases
  - Markets for long maturity
  - Other tree products related to public policy
  - Tree tenure, harvesting rights and exports control.

Strategies for adaptation

Keeping in view many causes, some modalities have been suggested to over come the problems. There must be strong linkage between researchers, industries and farmers. Hence the main objective is that the farmers should not suffer.

Collaboration between researchers extension and landholders

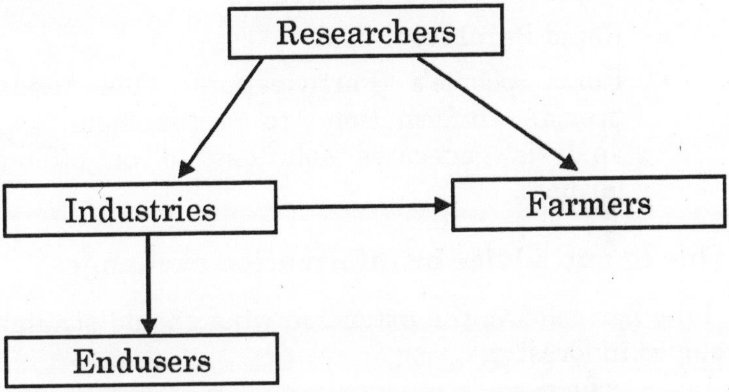

- **Identifying economic and other benefits**

Case studies providing objective analysis of economic, environmental and social impact of Farm Forestry/ Agroforestry at whole farm level should be used to inform landholders and other stakeholders of its potential benefits.

- **Diagnosis and Design**

The diagnosis and design approach encourage landholders ownership of the decision and also contribute more effectively to the development of a culture of continuous learning about forestry/Agroforestry.

- **Linear Top-Down Strategies**

In this, new technologies and knowledge are typically developed and validated by research scientists and that the task of extension agencies is to promote the adaptation of these technologies by farmers. It is also known as "target or blueprint oriented approaches."

Participatory "Bottom-up" strategies

The top-down approach must be avoided. What farmers need, accordingly technologies must be developed for time.

- Agro-ecosystem Analysis
- Rapid Assessment Procedure
- Rapid Rural Appraisal
- Rural people's Participation- This tends to provide information to researchers, whose analysis generates solutions to be offered to farmers.

- **One to one advice or information exchange:**

Like agriculture, the extension wing should strongly be developed in forestry.

- The Extension and Advisory Spectrum.
- Information access and technology transfer through television, radio, rural press, information sheets and booklets.
- Formal and structural education and training (field days, agroforestry expo, university courses, short courses) etc.

Success in Agroforestry / Farm Forestry

With the people's participation, assured market, buy back arrangement and high economic return, larger area has been brought under tree plantation. Different companies are promoting different tree species for different region. In the recent past, good success has been achieved in forestry sector. Till eighties, most of the paper industries were dependant on forest for their raw materials. But due to involvement of big industries, now 50% of raw material is produced by them.

WIMCO, Punjab, Uttar Pradesh	Poplar
ITC, Bhadrachalam, Paper Board, Ltd.	*Ecucalyptus, Casurina* and *Leucaena*
WEST COAST Paper mills Ltd.	*Acacia mangium* x *A. auriculiformis* - hybrid
Hindustan Paper Mills	Bamboo
J.K. Paper, Sripur Orient Paper, AP Papers	Subabul/*Leucaena leucocephala*
Rajasthan & Tamil Nadu	*Ailanthus excelsa* for veneer/ match sticks etc.

With their joint efforts

- 500 veneer unit established
- Rs.600 million generated annually
- 1.5 million man days / year generated
- 30 thousand tonnes of fuel wood available
- Price of industrial wood reduced by 50%
- High yielding Poplar clones (G3, G-48) *Eucalyptus* clones (MG-39 for irrigated and MG-28, 37 and 38) for rainfed developed.

Similarly, subabul made a spectacular comeback in Prakasam and Kirshna districts of A.P. for pulpwood production within a short rotation of 3-5 years. The district caters to the pulp requirement of a number of paper mills *viz.* J.K. Papers, Sirpur, ITC, Orient Paper, A.P. Paper, West Coast Paper Mills etc. 1996-1997. An estimate of ECO-Tech services indicated that in Prakasam district alone, subabul worth Rs.56 crores are produced annually (2538 ha area). ITC have started on improvement of subabul and developed high yielding clones *viz.* S-14, S-22 and S-24 which yield 67 tonnes/ha/yr (Haques, 2003). They have arranged buy back arrangement by fixing the following rates:

Rs. 650 / t gate price

Rs. 1200 / t for higher girth

Rs. 350 / t stumps after.

Under transfer of technology programme, CAZRI, Jodhpur has adopted 10 villages. To demonstrate the different technologies developed by CAZRI, as well as by the line department with the people's participation, following technologies have been demonstrated (Harsh and Tewari, 2003).

- Nursery technology (> 2 lacs seedlings).
- Tree plantation on boundaries and farm fields.
- Budding of high yielding ber varieties.
- Gum tapping from *A. senegal* (>Rs. 2 Lacs costing, gum inducer chemical has been sold).
- Adopted water harvesting device (Tank).
- Adopted high yielding crop varieties of crops.
- Pasture development.
- Livestock improvement etc.

Thus, it can be concluded that if forest based industries are developed, if they are more remunerative, then people's participation will be more and forest cover can be increased.

References

Black, A.W., Karen Forage and Fionnuala Frost 2000. Extensions and Advisory Strategies for Agroforestry. A report for RIRDC/LWRRDC/FWPRDC joint venture. Agroforestry Programme RIRDC Publication no.00/184.pp.1-29

Chandra, J.P., 2003. Role of forest based industries / plantation companies in development of Agroforestry pp. 305-309. *In: Agroforestry: Potentials and opportunities. Agrobios (India) and Indian society of Agroforestry.*

Haques, M.S., 2003. Emergence of Subabul (*Leucaena leucocephala*) as on industrial trees. pp. 211-318. In Agroforestry: *Potentials and opportunities. Agrobios (India) and Indian Society of India.*

Harsh, L.N. and Tewari, J.C. 2003. Agroforestry systems in hot arid regions with special references to value addition 159-170.

In Agroforestry: Potentials and opportunities. Agrobios (India) and Indian Society of Agroforestry.

FSI, 2003. State of Forest Report. Forest Survey of India (MoEF), Dehradun.

FOREST RESEARCH IN GUJARAT

B.K. Sinha, I.F.S.
Deputy Conservator of Forests, Publicity & Liaison Division,
Forest Research Institute, Gandhinagar, (Gujarat),
E-mail: dcfres@gujarat.gov.in

ABSTRACT

Research in Gujarat has covered experiments in forest areas, as well as, their practical application in the fields, including research on NWFP and agro-forestry. Present and future status, regarding yield capacity of the forest, can be calculated from the sample plots. In all, 16 sample plots, covering an area of 32.45 ha, have been established to study seven commercial species. To collect good quality seed and also to improve productivity of forests, a total 4107 Candidate Plus Trees (CPTs) of 40 species have been identified and marked. Clonal Multiplication Areas (CMA) of 19 species have been established with the aim to get continuous supply of scion material for grafting so as to produce good quality true to type planting material. Establishment of trials on tissue culture raised seedlings has also been done. A total of 69 gene bank plots for rare and threatened species have been raised at different research centers. Introduction trial plots of seven species, 70 provenance trial plots and 5 progeny trial plots have been created at different centres. For supply of good quality seeds for plantations, a total of 2 Seed Production Areas (SPA), 4 Seedling Seed Orchards (SSO), a total 81 Provenance Banks, a total of 140 Seed Stands were selected/identified. Research projects on Ecology and Biodiversity of the state, research on Bamboo, Bio-Diesel, VAM and Lac are some other projects, which have been taken up by the Gujarat Forest Department. To carry out extensive research activities, a total of 15 research centers in different agro-climatic zones have been established by the Forest Department.

On going activities, as discussed in this paper, will continue with higher technical inputs. The future research in the state will focus on tree improvement programme, propagation and multiplication of high quality bamboo, development of superior clones and advance in nursery technology. Developing appropriate silvicultural practices in Joint Forest Management areas, integrating the NWFP management with silvicultural operations are other focused area of the research.

Introduction

The state of Gujarat with a geographical area of 196022 sq. km. accounts for 5.98 percent area of the country. The per capita forest availability in Gujarat is only 0.05 ha. against the national average of 0.08 ha. The largest coastline; two of the three gulfs in the country, unique saline desert of the Ranns of Kachchh, vast grasslands, extensive thorn and dry deciduous forest in Saurashtra and North Gujarat, moist to semi-moist forests in South Gujarat and a large area under variety of wetlands, lakes and mountains make this state unique in many respects. Gujarat is a forest poor state yet it is rich in biodiversity. The forest area (Champion and Seth, 1968) of Gujarat State is 19113.0 sq. km. (i.e. 9.75%) of the geographical area but the actual forest cover area is 14,946.0 sq.km. (i.e. 7.63%). Gujarat is one of the few states in the country which has shown a continuous increasing trend in the forest cover since 1989. Most of the forest of Gujarat is dry deciduous to scrub type having very low productivity. The growing stock of Gujarat as assessed by Forest Survey of India (FSI, 2003) is 83.797 M.CMT with 56.06 CMT per ha.

Material and methods : Power Point Presentation.

Result and Discussion

In order to achieve long term goal of forestry research a set of objectives are assigned.

- Increase in productivity through selection, tree breeding and genetic improvement of planting stock.
- Adoption of advanced scientific techniques for plantation stock improvement.
- Conservation and restoration of gene-pool of plant species of rare, endangered, medicinal and ethno-botanical importance.
- Technology development for afforestation of drought prone, saline, barren and ecologically fragile areas through appropriate plantation technology and species.
- Introduction of economically and ecologically viable suitable exotics for strengthening social forestry.
- Improvement of NTFP through development of nursery and plantation technique.

The research work done till recently has been laying emphasis on studying the biological and ecological diversity of the areas, plantation technology for drought prone and ecologically fragile areas, improvement and adoption of latest nursery technology and propagation techniques of high yield varieties of important species.

Forest Research

1. Thrust Area

1.1 Infrastructure Development

A long term comprehensive research programme needs adequate infrastructure support. Under this programme, infrastructure has been given a high priority. The infrastructure includes:

1.1.1 Establishment of permanent research station

The focus on tree improvement, the long gestation period of tree species, various parameters which need continuous monitoring, improved nursery technology, field trials, centralized production, mass propagation etc. have necessitated the need of permanent research stations. Under this programme, nine such integrated stations have been established in different agro-climatic zones, each equipped with facilities of fencing, mist chamber, net house, vermi-culture unit, propagation unit, germination chamber, hardening ground, compost unit, and trial area etc. Details of research stations are given below (Table 1).

Table 1: Details of Agro-climatic Zone wise distribution of Research Stations

S. No.	Name of Centre	District	District Covered	Zone	Year of Establishment
1.	Rajpipla	Narmada	Dangs, Valsad, Navsari, Surat, Narmada, Bharuch	A	1981-1982
2.	Kheralu	Mehesana	Mehesana, Sabarkantha	D-2	1996-1997
3.	Anjar	Kachchh	Kachchh	E	1996-1997
4.	Bhavanagar	Bhavanagar	Bhavnagar, Amreli	G	1996-1997
5.	Godhara	Panchmahal	Panchmahal, Vadodara, Kheda, Dahod, Anand	C	1998-1999
6.	Gandhinagar	Gandhinagar	Gandhinagar, Ahmedabad	D-1	1999-2000
7.	Deesa	Banaskantha	Banaskantha Patan	F	1999-2000
8.	Junagadh	Junagadh	Junagadh, Porbandar	H	1999-2000
9.	Rajkot	Rajkot	Rajkot, Jamnagar, Surendrnagar	H	2003-2004

Under infrastructure development base research stations are established in all the agro-climatic zones, of the state but it is required to further strengthen these centers with other basic facilities.

1.1.2 Laboratories and Equipment

The field of seed technology and tissue culture etc, need fully equipped, laboratories. Under the programme, seed testing laboratories, soil testing laboratories, tissue culture laboratories, seed godowns etc. have been established along with the related equipments at Gandhinagar and Rajpipla (Anon, undated a, b). These laboratories require further upgradation particularly for identifying micro nutrients, plant fungal association, biofertilizers etc.

1.1.3 Research and Demonstration Centers

For disseminating the information related to field research, to public, for awareness and adoption, 19 R & D centers have been constructed under the field divisions. Future strengthening of these R & D centers are required. 19 centers one in each district have already been established. So it is proposed to establish another six new R & D centers, establishing one center in each district (Table 2).

Table 2. Research and Demonstration Centre Established in Field Division

No.	R & D Center	Division	District	Year of Establishment
1.	Chand-Tekri (Modasa)	S.K. (S)	Sabarkantha	1996-97
2.	Kheralu	Research, G'nagar	Mehsana	1996-97
3.	Bhavnagar	Bhavnagar	Bhavnagar	1996-97
4.	Kanjeta	Baria	Dahod	1997-98
5.	Bhavnath	Junagadh	Junagadh	1997-98
6.	Harij	Patan	Patan	1997-98

7.	Dhanol	S.F. Godhra	Godhra	1997-98
8.	Thorala	Jamnagar	Jamnagar	1998-99
9.	Piplaj	Chhota - Udepur	Vadodara	1998-99
10.	Dabkhal	Valsad (N)	Valsad	1998-99
11.	Zankhvav	Rajpipla (W)	Bharuch	1998-99
12.	Gandhinagar	Res. Gandhiangar.	Gandhinagar	1999-00
13.	Chotila	S'nagar (SF)	Surendranagar	1999-00
14.	Bhachau	Research, G'nagar	Kutch	1999-00
15.	Bilesvar	S.F., Bharuch	Bharuch	2000-01
16.	Ranifalia (Vansda)	Valsad (N)	Navsari	2000-01
17.	Waghai	Dangs (S)	Dangs	2001-02
18.	Kathlal	S.F., Nadiad	Kheda	2001-02
19.	Bodakdev	S.F., Ahmedabad	Ahmedabad	2001-02

2.0 TREE IMPROVEMENT STRATEGY

In view of the role of forestry in rural economy and social development, expectation thereof and the crucial role of research in providing the solution, the research focus under IFDP has been kept on tree improvement, emphasis on fast growing and high yield varieties, appropriate technology of mass production and plant propagation .

2.1 Tree Improvement Programme

To meet the diverse tree needs of the state, long term sustainable, qualitative and quantitative improvement in tree crop is imperative and accordingly a long term tree improvement programme has been formulated, along with its constituent components. This includes:

2.1.1 Identification of Superior-genetic Material

One hundred ten species of fruit, timber, and of NTFP importance have been identified. In the first stage 4107 CPT (Candidate Plus Tree) have been identified from the different agro-climatic zones, covering both forest and non-forest areas. Records and registers of the same are meticulously maintained. In the first phase, 40 species are identified for which different field trials have been started. The long term tree improvement plan prescribes provenance trials, provenance banks, progeny trials, seedling seed orchards etc. for the above identified tree species. Once the trials for above 23 species are over, further species will be identified for taking new trials.

2.1.2 Multilocational Tree Improvement Trials

As part of continuous long term programme, the provenance trials, progeny trials of superior selection have been initiated under the programme. Uptill now provenance trial of *Ailanthus, Casurina, Eucalyptus*, Sisu (*Dalbergia sisso*), Guggal (*Commiphora mukul*), Teak (*Tectona grandis*), Neem (*Azadirachta indica*), Mahua (*Madhuca indica*), Harde (*Terminalia chebula*), Baheda (*Terminalia bellerica*), Amla (*Emblica officinalis*), Aritha (*Sapindus* sp.), have been taken at various research stations.

Progeny trial of Khair (*Acacia catechu*), *Eucalyptus*, Sisu (*Dalbergia sisso*), *Causurina*, *Ailanthus* and Teak (*Tectona grandis*) have also been taken.

2.1.3 Seed Orchards

For further breeding and better seed source, clonal seed and seedling seed orchards have been established for *Casurina* and Sisu (*Dalbergia sisso*), Neem (*Azadirachta indica*), Ardu (*Ailanthus*), Mahua (*Madhuca indica*) and Baheda (*Terminalia bellerica*), etc. So far, 122 clonal orchards of 18 species and 168 seedling seed orchards of 33 species have been established.

2.1.4 Gene pool Preservation Plots

Endangered and rare plant species have been identified and effort has been made to establish the gene bank. The gene pool preservation plots of Gum Kadaya (*Sterculia urens*), Guggal (*Commiphora mukul*), Billi (*Aegle marmelos*), Rayan (*Manilkara hexandra*), Charoli (*Buchanania lanzan*) and Sisham (*Dalbergia latifolia*) has been established. So far, 35 gene pool preservation plots of 10 species have been established at different research stations.

2.2 Clonal Propagation Programme

For increasing the productivity in a short period with reference to specific trials, focus has been kept on clonal propagation programme. This includes

2.2.1 Selection of superior Clones

Thus, superior clones have been selected from forest and non forest areas. Besides this superior clones have been procured from other state as well.

S.No.	Species	No. of clones	Areas
1.	*Eucalyptus*	78 clones	A.P., Tamilnadu, Gujarat
2.	*Casurina*	17 clones	Tamilnadu
3.	*Dalbergia*	55 clones	U.P., Rajasthan, Gujarat
4.	*Tamarindus*	19 clones	A.P.
5.	*Sitaphal*	3 clones	Gujarat
6.	*Zizyphus* sp. (Bordi)	3 clones	Gujarat
7	*Emblica officinalis* (Amla)	3 clones	Gujarat

2.2.2 Clonal Trial

The superior clones are to be tested for their performance and genetic stability. Therefore, an elaborate programme of clonal test has been started. Accordingly, clonal trails of *Eucalyptus*, *Casurina*, Sisu (*Dalbergia sisso*), Pilu (*Salvadora persica*) selected are being established in different research stations.

2.2.3 Establishment of Clonal Multiplication Garden

To make stock for clonal propagation readily available, clonal multiplication gardens have been established at research stations. The species include Nilgiri (*Eucalyptus*), Amla (*Emblica officinalis*), *Casurina*, Sisu (*Dalbergia sisso*), Khati imli (*Tamarindus*), Harde (*Terminalia chebula*), Baheda (*Terminalia bellerica*) etc. So far, 45 clonal multiplication gardens of 19 species have been established.

2.3 Production of Improved Planting Stock

In order to ensure that superior genetic selections are used in field, the research wing has focused on mass scale production also.

2.3.1 Production of Root Trainer Plants from CPT

Thus, seed of superior selection have been used to produce genetically superior plant. The root trainer technology helps in developing profuse root system for increasing root surface area considerably which in turn helps in establishing the seedlings and better nutrient absorption. Up till now, approximately 32.5 lakhs seedlings have been produced by research wing and have been distributed to territorial and extension divisions of department.

2.3.2 Clonal Plants

Superior clones of Nilgiri (*Eucalyptus*), *Casurina*, Sisu (*Dalbergia sisso*), Pilu (*Salvadora persica*) have been produced for field plantation. So far, more than 1.24 lakhs plants have been produced.

2.3.3 Production of Grafted Plants

Forest fruit yielding plants have high economic value especially in rural economy. Therefore, the research focus has been to standardize the grafting techniques for different species. Over the period, grafting has been standardized for

Amla (*Emblica officinalis*), Harde (*Terminalia chebula*), Baheda (*Terminalia bellerica*), Aritha (*Sapindus laurifolius*), Sitaphal (*Annona squamosa*), Mahua (*Madhuca indica*), Khati imli (*Tamarindus indica*) etc. Uptill now 8.51 lakhs grafted plants have produced and distributed to farmers and the territorial wing of department. Amla (*Emblica officinalis*) is occupying major portion in grafted seedlings and farmers have actively adopted it for horticulture farming.

2.4 Micropopagation

For future use and application of tissue culture, the standard protocol for various forest tree species is being developed. Different Institutes / Universities were assigned the project. The species includes:

1) *Diospyrus melanoxylum*	6) *Commiphora vitis*
2) *Ailanthus excelsa*	7) *Sapindus emarginatus*
3) *Terminalia bellirica*	8) *Dalbergia latifolia*
4) *Terminalia chebula*	9) *Acacia catechu*
5) *Tecomella undulata*	10) *Buchnania lanzen*

So far, mass propagation of safed musli (*Chlorophytum tuberosum*) medicinal plant of economic importance has been undertaken. Up till now, 10,000 plants have been produced and are being used for field trial. The protocols developed for above mentioned species are put to standardization at Rajpipla and Gandhinagar research stations. Species multiplication has been achieved successfully but could not be established in the field. Further research is needed for which trained personal of research fellows on contract basis are required.

3. Option of New Technology

In the era of fast changing world, the conventional forestry needs to be substituted by new proven technologies of nursery, plant production and plant propagation which provide distinct technical and economic advantages. Under IFDP, programme efforts have been undertaken to adopt tested and proven technologies for mass scale use in field.

3.1 Green House Technology

Controlled environmental conditions provide definite advantages in seedling production and more so in vegetative propagation. It reduces the nursery period and affords capabilities of standardization. Accordingly, this technology has been adopted. Up till now 9 green houses/mist chamber have been erected and made operational under the research wing. For its use by field staff, 19 green net house units have been constructed in various divisions. It is planned to construct 6 more units in central and north Gujarat.

3.2 Root Trainer Technology

Root trainer raised seedlings offer distinct advantage over poly bag seedlings in terms of survival capabilities, nursery raising period, transportation etc. and gradual substitution of conventional poly bag method with root trainers is needed. Therefore, under this programme root trainer technology has been adopted and standardized by research wing. Up till now approximately 24,00,000 root trainer seedlings have been produced and tried in field by Dang, Vyara, Rajpipla, Godhra, Valsad, Gandhinagar, Mehesana, Banaskantha etc. divisions.

3.3 Compost & Vermi-Compost Producing Technology

Manure production by compost & vermiculture compost is an eco-friendly technology that uses fertile bio-degradable material and enables recycling of waste. It reduces the use and consumption of chemical fertilizers. Looking to the fertilizer requirement of forestry, adoption of this technology is needed. Up till now, 10 such units have been established and is being used by research wing. Gradually, it is being provided to field divisions as well. Awareness campaign will also be organized to educate the local farmers living in and around the center.

3.4 Clonal Technology

Though traditionally, the agronomy and forestry has relied upon seedling population, all the more, during the recent past, the concept of clonal forestry has become a major strategy for production of specific tree crops with desirable traits.

Clonal technology involving clonal selection and clonal propagation is revolutionary method for achieving quicker genetic gains in terms of productivity and yield over short periods. Several countries have adopted it in a big way. Therefore, during the project period this technology is being adopted for its future use. Thus, clonal selection, clonal trials, clonal propagation and clonal banks have been undertaken by research wing in different species like Nilgiri (*Eucalyptus*), Saru (*Casurina equisetafolia*), Sisu (*Dalbergia Sisso*), Pilu (*Salvadora persica*), Guggal (*Commiphora mukul*) etc. were distributed to farmers. Approximately 45 ha. area under clonal multiplication area and scion banks have been developed. During 2004-05 approximately 74,000 clonal seedlings have been raised and distributed to the farmers under Van Mahotsava.

3.5 Adoption and Use of Bio-Technology

The productivity in field of agriculture and horticulture has been increased many times by using modern techniques. Recent examples of successful application of these techniques in forestry by various countries provide a possible direction in this field. Both for the purpose of technical up-gradation and also to take the advantages of these latest technologies, the project laid emphasis on adoption and use of biotechnology. Keeping future objective of research, it is aimed to focus on tissue culture technology and bio-fertilizer.

3.6 Tissue Culture Technology

The modern technique of tissue culture provide the ability of large scale multiplication of superior genetic material, especially, the species having biological problems of germination and regeneration. It is also the basic technology for genetic engineering. Therefore, under the project, efforts have been made to adopt this technology and create a base for future use and adoption. Accordingly, fully operational tissue culture lab has been established and protocols are being established for different species like Teak (*Tectona grandis*), Safed musli (*Chlorophytum tuberosum*), Baheda (*Terminalia bellitica*), Harde (*Terminalia chebula*), Ragat rohido (*Tecomella undulata*) etc. Tissue culture is a costly technique in the initial stages and cannot be used to substitute the macro propagation methods for large-scale production. In order to keep the focus clear, only those species have been taken up for mass scale production, which have extraordinary characters of yield or adaptability, which have poor seed production or have problems of natural regeneration or which have very high economic value. It is planned to carry out research on Bamboo, *Ailanthus*, *Eucalyptus* and medicinal plants in future.

3.7 Adopting the use of Vam and Bio-Fertilizer

It has been decided to introduce VAM and bio-fertilizer use technology at different research centers with the help AFRI, Jodhpur (Rajasthan). But, it is only at an experimental level. Now, it is proposed to take up these technologies on a large scale. These technologies are very relevant for raising healthy, disease resistant and hardy plantations with high survival percentage. A full fledged scheme for adopting VAM is going to be implemented very soon. On a pilot project basis, it has been tried at Basan research station and results were encouraging. After laboratory testing, it has been found that inoculations were successful for six different strains of VAM. It is provisioned that through this scheme, staff will get itself familiarized with the latest techniques. It shall also provide the scope and opportunities of testing and using this highly sophisticated technology in the department, may be on an experimental scale. It shall also provide scope of establishing technical collaborative projects and assignments with the institutes and agencies having infrastructure and professional expertise in this field. For isolating species-specific strains of VAM and bio-fertilizers and its testing for multiplication, supply and use in the fields for raising healthy, disease resistant and hardy plant seedlings with high survival percentage, it is proposed to start a new project on large scale.

4. Programme of Seed Improvement & Development

Seeds are the genetic storehouse of nature and the genetic diversity found in the seeds, provides the genes from which the plant life covers the most of the land surface in all of its environmental variations. The seed production process, involving crossing between different parents, provides the base of genetic variability for various traits, which offers the opportunity of natural selection and evolution. Seed selection is an effective method for genetic improvement and has been used successfully in domestication of various agricultural and horticulture crops. Mass propagation by seeds is an efficient and economic method of propagation. Therefore, for the purpose of securing quick genetic gains, in terms of productivity, disease resistance, wide adaptability, for the purpose of long term genetic improvement, for the purpose of crop domestication by specific breeding program, and to utilize these genetic gains in field by efficient and economic method, a full emphasis has been given on seed development and improvement program.

The seed development program has the present capacity to supply only 40-50% of the total seed requirement of the state. Despite the heavy need of plantation stock, both by the department, as well by the people; the seed program has received low priority, in terms of technical and financial inputs over several plan periods. Though seed production areas for 34 species have been established, however related operations in the form of continued program are lacking, leaving a situation where majority of seed requirement is fulfilled through local collection, with little control over the quality. The testing and storage facilities are also limited.

4.1. Improvement of existing Seed Stands

Since last many years, 140 seed stands of 34 species have been established. Due to paucity of funds, necessary improvement operations have not been undertaken, leaving low output from these areas. In the present project, the existing seed stands shall be assessed afresh and unsuitable ones shall be rejected and the screened and finalized seed stands shall be used for the seed supply, to increase the production and quality of seeds.

Though old stands cover many species, now the emphasis will be on only target species. These are Teak (*Tectona grandis*), Harde (*Terminalia chebula*), Baheda (*Terminalia bellitica*), Saru (*Casurina equisetifolia*), Pilu (*Salvadora persica*), Guggal (*Commiphora mukul*), Khair (*Acacia catechu*), Sadad (*Terminalia tomentosa*), Mahua (*Madhuca indica*), Neem (*Azadirachta indica*), Babool (*Acacia nilotica)*, Nilgiri (*Eucalyptus*), Sisu (*Dalbergia sisso*).

4.2. Establishment of New Seed Stands

In order to ensure the supply of superior seeds of target species and also to ensure infusion of fresh genetic material, the project envisages the establishment of new seed stands. Under the scheme, while more area shall be brought under seed stands for the species already having seed stands, at the same time, seed stands shall also be established for the target species, not covered earlier.

The seed stand area shall be increased for Harde (*Terminalia chebula*), Baheda (*Terminalia bellerica*), Rayan (*Manilkara hexandra*), Guggal (*Commiphora mukul*), Pilu (*Salvadora persia*), Sadad (*Terminalia tomentosa*), Mahua (*Madhuca indica*), Neem (*Azadirachta indica*), Nilgiri (*Eucalyptus*), and fresh stands shall be established for Sevan (*Ougenia ogeinensis*), Haldu (*Adina cordifolia*), Shisham (*Dalbergia latifolia*), Ragat rohido (*Tecomella undulata*).

5. Co-ordination and Linkage with other Agencies

Various independent agencies, individuals, professionals, government departments posses the infrastructure and technical expertise which can be of assistance in forestry research. Therefore, a healthy co-ordination and linkage is needed. Under IFDP programme, this has been established and various collaborative research projects have been undertaken (Table 3.).

Table 3: The number of research projects under taken

S. No.	Project field	No. of Studies
1.	Biodiversity	6
2.	Propagation techniques	9
3.	Ecological Studies	6
4.	Wildlife Studies	3
5.	Socio Economic Studies	6
6.	Management related studies	8

6. Research Linkage with Field

In order to ensure that the achievement of research wing reaches to the people through out the department in large scale, the capacity building is done through imparting training to field staff in tree improvement and nursery techniques. As a part of research linkage to field, the superior seedling raised by the research wing is supplied to department for raising plantations. Similarly, plants raised from CPT seeds, Clonal plants and grafted plants have been supplied to department for distribution to farmers.

Adoption of technology and its standardization at the base research station and its further transfer at field level is an important aspect of forestry research. For information dissemination, different tools are adopted

6.1 Publishing Technical Reports

Different techniques which are cost effective, time saving, efficient and productive are developed at research stations by conducting experiments. These findings are released in the form of technical reports and are sent to all the divisions for their utilization in field. Technical reports on soil solarisation, sprinkler system in nursery, seed germination, grafting technique of amla etc. were circulated and many more are in the process of finalisation.

6.2 Khedut / Farmers Shibir

As a part of information dissemination and transfer of technology from research to field for adoption, farmer camps were undertaken to draw maximum advantage of the facilities so created. It may be mentioned that the response of these camps conducted at research centers has been very encouraging and as a result, many demonstration plots from superior quality planting stock have been developed by farmers. In last three years, 277 farmers shibir/ camps have been organized on these research centers in which 14286 farmers have participated. During these camps, mostly emphasis is given on agro-forestry models, grafting technique and its commercial viability and vermi-compost.

6.3 Staff Training

It is imperative to impart training to field staff before launching the adoption programme for new techniques and technology. Keeping this view into consideration, extensive field staff trainings were organized on research centers. The selected staff from territorial and social forestry wing are trained for root trainer technology, CPT selection, seed collection, grafting technology, vermi-compost, and are exposed to modern technologies of planting stock development such as tissue culture, micro-propagation and clonal seedlings development. In the last two years, more than one thousand staff of different cadre has been trained. The trained staff is working as a human resource for technology transfer from lab to field.

7. References

FSI, 2003. State of Forest Report 2003. Forest Survey of India, Dehra Dun.

Anon undated a. Official records of Forest Research Division, Gandhinagar, Gujarat Forest Department.

Anon undated b. Official records of Silviculture Division, Rajpipla, Gujarat Forest Department.

Champion H.G. and Seth S.K., 1968. A revised survey of the forest types of India, Manager of Publications, Government of India, New Delhi.

EXTENSION STRATEGIES FOR ENHANCED USAGE OF LESSER-KNOWN WOOD

S.H. Jain and R. Arya

NWFP Division, Arid Forest Research Institute, Jodhpur

ABSTRACT

It has not been realized that preservative treatment makes timber and timber based products durable and thereby cutting down the repeated use of timber, thus helping the conservation of natural resource. Treatment makes timber resistant to biological attack and immune to termite and borers. A common man does not know about treated timber. Public awareness is necessary for the growth in treated timber usage. Even if few people know about treated timber, may not be knowing right preservative to be used and right method of treatment. Educating people about treated timber is a huge task. Therefore, the only way to educated people is through Government/ Research Institutes initiatives. Government has taken lot of efforts to popularize consumption of milk and eggs, usage of CNG and biodiesel etc. Similarly, Ministry of Environment and Forests should take initiative to promote usage of treated timber.

At AFRI a treatment schedule has been developed for 3 plantations grown lesser-known tree species *viz. Acacia tortilis, Prosopis juliflora* and *Prosopis cineraria.* The treatment is very effective in case of *A. tortilis* and *P. juliflora* with very good finishing properties whereas *P. cineraria* had poor finishing properties. State Forest Department, Rajasthan has taken up massive scale plantation of *A. tortilis,* showing promising results in establishment on sandy area. The wood of *A. tortilis* is highly susceptible to insects and pests but preservative treatment makes timber resistant to biological attack. *P. juliflora* grown as weed and used for fuel wood may be encouraged to be grown as clear bole tree by pruning side branches which do serve both purposes of timber, as well as, firewood. This technique can be popularized and extended to handicraft industry, sawmills, carpenters, artisans and other allied small-scale industry. This may be extended by way of exhibiting products in fairs/ melas, organizing demonstration-cum-training programmes for group of industrialists and other end users.

Sandal (*Santalum album*), is another important lesser known species in this region. This specie existing naturally in semi-arid region is also successfully grown in arid region by proper management like providing suitable host plant, shade and irrigation. It is proved beyond doubt that tree contains heartwood and oil in Chittorgarh region. This is one of the costliest woods having high demand in the market by artisans and handicraft industry. Few farmers are already growing it on boundaries and trenches of agricultural land. Cultivation of this specie may be popularized by organizing training and demonstration programmes to staff of forest department, farmers, artisans, NGOs, etc.

Introduction

Timber is a major component of traditional handicrafts. The role of handicraft industry is very significant in Rajasthan. It offers livelihood and employment to many. The scarcity of traditional timbers and their price escalation poses problems to these industries. Source of supply of timber is from neighboring states. Adoption of appropriate post harvest technologies is required for efficient utilization of lesser known secondary species.

Utilization of three lesser-known secondary timbers *viz. Acacia tortilis, Prosopis juliflora and Prosopis cineraria* were made possible through developing seasoning and preservative treatment schedule. The wood properties are comparable to shisham. In order to ensure continuous timber supply to industries and general human needs, several fast growing species particularly *A. tortilis* has been planted for sand dune stabilization by Rajasthan Forest Department. These species are harvested on short rotation, therefore, have higher percentage of sapwood and growth stresses. They pose problem of deterioration due to splitting and attack by insect pests, thus need modified technology and special care in processing.

Extension is a tool to transform the research results transferred into useful information to end-users. The principal objective of an extension process is to enable people to use the developed technology and to promote social, cultural, recreational, intellectual and spiritual life, and thence improve quality of life of the common people. Directorate of Extension will be responsible for planning coordination and supervision. Execution of various programs will be the responsibility of extension divisions of the Institutes. Scientists who have developed technologies will be involved from planning to implementation stage (Anon 2000).

Material and Methods

Under the ICFRE plan, the project was "Studies on post harvest technologies of non-traditional, under exploited locally available timber species for suitability to handicraft and other small scale industries" under taken during the year 2003. The infrastructure of the private wood treatment plant was used.

Wood of three plantation grown lesser known species *viz. A. tortilis, P. cineraria* and *P. juliflora* were obtained by felling trees (80-90 cm girth) from experimental field of Arid Forest Research Institute, Jodhpur. Wood logs were sawn into required size. Sawn wood pieces were then impregnated with preservative chemicals, copper chrome arsenic (CCA) 2% and chloropyriphos 2% solution in pressure treatment plant treated for 3 hours at 80 psi. Treated wood had been seasoned in steam generated seasoning kiln till the moisture content was brought down to 10-12%, further seasoned in sunlight and shade. Seasoned and preservative treated wood had been used to make value added products like sofa set and utility box. Few pieces of wood were kept untreated for observation and comparative study of deterioration.

Results and discussion

The combination of preservative chemicals Copper Chrome Arsenic (CCA) and chloropyriphos were found effective against insect pest attack. Wood from A. *tortilis, P. cineraria and P. juliflora* treated with these chemicals enhanced the resistance to fungal and insect pest infection and there by increased durability and shelf life compared to control samples. The deterioration of wood commenced within 2 months of felling and deteriorated to the extent of 50% within a year. Where as treated wood are showing resistance. Untreated wood of *P. juliflora* was not infected with any decaying agents.

Value added product like sofa set, utility box and pen/pencil stand have been made from treated wood of these species to study the performance of wood. Prelimanary work on preservation and seasoning of timber of three species *A. tortilis*, *P. cineraria* and *P. juliflora* have shown promising results, further study, observations and different combinations of eco friendly preservatives can be tried to consolidate the schedule of treatment and shelf life enhancement of these woods. This work is very important work for north-west region of India as the Rajasthan Forest Department has taken up large- scale (0.5 millin ha.) plantation of *A. tortilis* for sand dune stabilization and roadside plantation. *P. juliflora* spreads in vast area as weed; villagers use branches for fire wood purpose. Wood is resistant to pests, insects and fungus, but treatment of preservative further strengthen the resistance for long durability. This specie need be popularized to be grown as clear bole tree to get required timber. *P. cineraria* is having limited scope as timber, because it is not allowed to be felled in this region and has other multiple utilizier as fodder and food.

Timber of plantation grown *Acacia nilotica* is already being accepted and used by many wood industries in India. Utilization of *A. tortilis* and *P. juliflora* with appropriate preservative treatment and seasoning can substitute the traditionally used wood. Application of this technique on utilization of plantation grown timbers can reduce pressure to some extent on supply of traditionally used timber species like teak and shisham, thus, the conservation of natural resource.

Table : Wood properties of three species

Species	Resistance to wood decaying agents	CCA and Chloropyriphos	Mech Properties	Finishing properties	Utility
A.tortilis	Highly susceptible	Effective	Good	Very good	Handicraft industry
P. juliflora	Resistant	Effective	Very good	Very good	- do -
P.cineraria	Highly susceptible	Effective	Poor	Poor	Agricultural implements

Extension activities

The extension work on these technologies developed were done with the cooperation of Extension Division, AFRI. One of the most important method was displaying exhibits of these products at exhibitions/fairs/melas/seminars/ workshops and conferences with detailed technology hand outs. The target groups are handicraft indusrties, sawmill owners, carpentars, artisans, etc.

Extension strategies

- Public awareness is necessary for the growth in treated timber usage.
- People must be educated through Government/ Research Institutes' initiatives.

Advantages of preservative treatment and seasoning of secondary plantation species

- Plantation grown secondary timbers are more susceptible to insect/pests and wood rot fungus.
- Treatment makes timber resistant to biological attack and immune to termite and borers.
- Treatment makes timber and timber based products more durable.

Target groups

- Handicraft industries
- Small and big wood based saw mills
- Entrepreneurs
- Carpenters
- Artisans
- House builders
- Common man

Awareness campaign

Public awareness about technologies and their advantages is to be created using

- Mass media: News paper, TV and radio.
- Print media: Booklets, bulletins, pamphlets, handouts and posters.
- Slides in cinema hall.

Conclusion

Training and demonstration

- Treatment technique should be popularized by way of exhibiting products in fairs/melas and organizing demonstration-cum-training programs for group of industrialists and other end users.

- Linkage between Forest Department and Wood Industries Association.

Government policy

- In most of developed countries, there are strict regulations to use only treated timber.
- Supreme Court has advised the Forest Department of Andaman and Nicobar Islands to sell only treated timber. Such steps in other states would also enhance the usage of treated wood.

Literature consulted

Anonymous 2000. Forestry research extension programme ICFRE, Media division Directorate of extension, ICFRE Dehradun.

Anonymous 1964. Timber seasoning: pole drying. 1963, Extr. from Rep. For. Prod. August. 1963/64.

JHEA. 2002. 4th Annual Report of Jodhpur Handicraft Exporters Association 2003-2004.

Nagaveni, H.C; Ananatha Padmanabha, H.S., Vijayalakshmi, G.; Sharma, M.N. and Shankaranarayana, K.H. 1998. Protection of rubber wood with modified creosote. Wood preservation doc. No. IRG, IWP 98-30,165. Netherlands 14-19 June, IRG Secretariat, Stockholm, Swedan.

Pandey, C.N; Arora, K.L; Kanoji, H.C. and Kambo, A.S. 2000. Suitability of *Ailanthus excelsa* for pencil making. Wood Seasoning Discipline, Forest Products Division, Forest Research Institute, Dehra Dun; India. *Journal of the Timber Development-Association of India*, 45: 3-4, 16-20, 8.

Ranjana Arya; Jain, S.H. and Chaudhuri, K.K. 2004. Status of lesser known trees in Rajasthan for their potential uses. *Myforest*, 40(4): 421-431.

Sanderson, M.S. and Harris, P.J.C. 1991. A quick guide to useful nitrogen fixing trees from around the world. NFT Highlights Winrock International Arkansas 72110-9370, USA NFTA 91-04, July 1991.

Shankaranarayana, K.H and Kumar, Vineet 1995. Structure and Industrial application of TEMD. *Research and Industry* 140: 293-294.

Jain, S.H.; Kumar, H.; Ranjana Arya and Chaudhuri, K.K. 2003. Woodcraft industry in Rajasthan – A challenging opportunity for Wood Preservation. National workshop on wood presrvation in India: Opportunities, challenges and strategies, 20th – 21st Oct 2003 at IWST, Bangalore.

Tewari; J.C.; Harris, P.J.C.; Harsh, L.N.; Cadoret, K. and Pasiecznik, N.M. 2000. Managing *Prosopis juliflora* (Vilayati babul). A Technical manual.

FORESTS VERSUS ORANS AND GAUCHARS

S.M. Mohnot and Kuldeep Singh
The School of Desert Sciences, 109, Nehru Park,
Jodhpur-342 003
Email: sdsjodhpur@sancharnet.in

ABSTRACT

The Great Indian Desert or Thar is world's most populated hot desert (density 84-90 person per sq. km.). No desert in the world is so people supportive as Thar. It is probably because of the mechanism developed by people as science of survival to face the vagaries of climate, inhospitalities and shortages. In over 55 years of rigorous planning after independence and huge layout of funds, the great Indian is still reeling under recurrent droughts wherein plight of people is visible in every drought. A considerable proportion of native population belonging to different castes and creeds but predominantly Scheduled Castes, Scheduled Tribes and indigenous people who are marginalized group of people. They suffer because of non-availability of natural resource support as well as inaccessibility to resource in the Common Property Resources in villages they belong to. While the human population is mounting, the livestock quality and quantity is decreasing leading to poor economic turnout. We must understand that desert's economy is livestock driven and a biomass animal balance has to be understood in its totality. The biomass support repositories, the orans (village forests) and gauchars (village pastures) are most important village institutions being the storehouses of desert's biodiversity, which support both ecology, and economy of the desert. Orans are sacred silvipasture regimes while gauchars are annual nutrition banks. Orans and gauchars along with other village institutions constitute about 5.8% of the total land as against about 1.8% of forests in the Great Indian Desert. The obvious importance of these institutions call for massive inputs of science and technology to revive these life-supporting institutions. In western Rajasthan, the orans or sacred groves varies from few hundred bighas to as large as 55,000 bighas. This is perhaps the largest oran in desert located in Barasingo-ka-Bada in Phalodi, Jodhpur, followed by oran at Bhadariyaji, Jaisalmer (42,000 bighas). Interestingly each village has got a small to large oran in whole of the desert. Most of the orans support 2 to 4 villages, some 5 villages, few up to 30 villages and 2-3 orans support as many as 100 villages. This suggests the importance of orans in desert. Likewise, in Thar about one-lac hectares of land in 12 desert districts represent bad to good gauchars. But, in general the condition of gauchars is poor. These lands are loosing their character and productivity every day because of neglect by the community and the State both.

We have worked out large number of factors for this decline, which may be physical, physiological, social and religious in nature. But, the most important factor appears to be inordinate delay in taking the appropriate technology and research finds to the people besides lack of political will and firm decision for redressal of desert biodiversity. A comprehensive plan for the revival of village institutions is needed. This must be addressed urgently for the good of rural masses who are either partially or fully dependent on natural resource for their survival as forest produce or produce of CPR's.

I. Introduction

The Great Indian Desert or Thar is world's most populated hot desert (density 84-90 persons per sq. km.). No desert in the world is so supportive as Thar. It is probably because of mechanisms developed by people in the form of science of survival where people can face the vagaries of climate, inhospitalities and shortages to survive. It is a living example of traditional wisdom and desert ethos still practiced, improvised and carried forward in the interior of the countryside by desert in-habitants in general and by rural poor for subsistence in particular. Interestingly, the so-called new science and technology has still not reached in these areas for variety of reasons. In over 50 years of rigorous planning following independence, and huge layout of funds, the Great Indian Desert is reeling under recurrent droughts. The plight of people can be seen in every drought. A considerable proportion of native population which belong to different castes and creeds predominantly of Scheduled Castes, Scheduled Tribes and indigenous people, who by and large, are marginalized group of people suffer because of declining resource in community institutions like 'orans' and 'gauchars'. The non-availability of resource support and at times their inaccessibility to resource in the Common Property Resources (CPR) is posing serious problem to rural poor due to inequity. While the human population is mounting, the livestock quality and quantity is decreasing, leading to poor economic turnout. We must understand that desert's economy is livestock driven and a biomass-livestock balance has to be understood. The biomass support repositories, 'orans' (village forests) and 'gauchars' (village pastures) are most important village institutions. They are the storehouses of desert's biodiversity, which support both ecology and economy of the desert. Orans are sacred silvipasture regimes while gauchars are annual nutrition banks. Orans and gauchars alongwith other village institutions constitute about 5.8% of the total land as against about 1.8% of forests in the Great Indian Desert. The obvious importance of these institutions call for massive inputs of science and technology to revive these life-supporting institutions. In western Rajasthan, the orans or sacred groves vary from few hundred bighas to as large as 55,000 bighas. The largest oran is Barasingo-ka-Bada in Phalodi, Jodhpur, followed by Bhadariyaji, Jaisalmer (42,000 bighas). Interestingly, each village is provided with a small to large oran in whole of the desert. Most of the orans support 2 to 4 villages, some 5 villages, but few up to

30 villages and 2-3 orans support as many as 100 villages. This suggests the importance of orans in desert. Likewise, in Thar about one-lac hectares of land in 12 desert districts represent bad to good gauchars. But, in general the condition of gauchars is poor. These lands are loosing their character and productivity every day because of neglect by the community and the state both.

We have worked out large number of factors for this decline, which may be physical, physiological, social and religious in nature. But, the most important factor appears to be inordinate delay in taking the appropriate technology and research finds to the people besides lack of political will and firm decision for redressal of desert biodiversity. A comprehensive plan for the revival of village institutions is needed. This must be addressed urgently for the good of rural masses who are either partially or fully dependent on natural resource for their survival.

In this paper, we have tried to deal dynamics of 'orans' and 'gauchars' in addition to 'agors'. The three together constitute very important trio called 'AGO'. We have dealt various facets of AGO from Bikaner, Barmer, Jaisalmer, Jodhpur and Jalore as examples. Tried to relate them with rainfall and livestock available and have taken clues from societal code of conduct with suggestions to improvise these institutions.

II. Problems of Thar Desert

In about one-lac years of Thar's existence, it had seen many geological and climatic upheavals through which it has survived. This unique but stressed eco-region cover about 0.32 million sq. km. of Thar. The inhospitable climate, extremes of temperature, deficit and erratic rains, lack of soil moisture, dry and arid conditions, absence of perennial rivers, nutrient deficient soils, vast tract of sand dunes and dust storms in hot summers make life difficult in the desert. Rains often fail leading to drought and famines keeping a considerable section of population under stress, which has to survive below poverty line. A region which used to fight the drought by adopting peoples science, traditional coping methods and endurance have become helpless due to poor biomass turnover or poor *'zamana'* as a result of droughts. On an average, Thar witness 6 drought years, 2 good years and 2 average years in a decade. Thus, only 2 are *zamana* years and remaining 2 are enough to carry on while the remaining 6 years are drought or deficit years.

In spite of such a precarious situation, the people of Thar have learnt to live with droughts by following such mechanism like collection, conservation and judicious use of rain water and furthering ground cover to augment fodder production. *'Gauchars'* and *'Orans'* are rich biodiversity zones, help conserve germ-plasm for posterity. This is evidenced by species richness, genetic variation and biological diversity in Thar. The *'Agor'* is yet another important village institution which collect and channel rain water to water bodies. The *agor-gaucher-oran* trio is a repository of biological diversity and need special attention.

III. Declining Trend of *Gauchar-Oran* Production

Thar possesses over 900 plant species, 85 of which are grass species. Many of these are potential palatable species supporting livestock population till recently. Over the decades, orans and gauchars had been the major source of free fodder. Forests in desert are around 1.8% of land while *agors-gauchars-orans* together constitute about 5.6% of the total land. The later is thus very important resource. It however, needs upkeep, maintenance and improvement to enhance productivity to meet demand of fodder, firewood, fencing material, thatch, herbal plants, etc.

IV. AGO and Societal Code of Conduct

The AGO have survived for centuries in adverse climatic conditions primarily because of people's faith, reverence and protection, they give to village institutions. It is perhaps the major factor for AGO's survival in desert. But, with increasing population pressure, demand of land for variety of works, increasing agriculture and commercialization, the protection enjoyed by AGO from village community is eroding.

AGO is a simple set of village institution with clear spatial stratification. Since the system is operating successfully for centuries without any problem for its use, physical and biological status, productivity, sanctity and ownership, it must be revived.

V. Gauchars (village grazing ground)

The role of *gauchars* in maintaining Thar's economy has been enormous till 4 decades ago. From 1960 with the onset of intensive agriculture and tube-well culture, the decline of *gauchars* started. In the process, some 400-year-old culture and tradition of *gauchars* in the desert started decaying. A general impression of present day *gauchars* in Thar can be made out from the following profile

General Status of *Gauchars* in Thar Desert

Age of Gauchars	> 100 years < 400 years
Size	Few hundred to 40,000 bighas
Founders	Philanthropists, Community Organisations, etc.
Purpose	Essentially Cattle Grazing
Control	People-Society-Community
Present Status	Decaying
Soil Profile	Poor
Vegetation	Open Scrub: Herbs, shrubs and grasses, some annuals.
Plant species	Highly degraded: Perennial: >17; Ephemerals: > 60
Pressure Period	Kharif, Rabi and Drought

Upkeep and Maintenance	Lacking
Scope of Intervention	Enormous

As per Government Revenue Records, Thar Desert represents about 0.85 lac hectares of land under *gauchars* in 12 desert districts. This provides valuable support to livestock economy. Interestingly, the three core districts of Thar *viz.* Jodhpur, Jaisalmer and Barmer constitute about 50% of total *gauchars* of the Great Indian Desert on which about 27% of Thar's total livestock depends. The inter-relationship of rainfall, livestock and size of *gauchars* for 11 desert districts is given in table below.

District wise *Gauchars* of Thar in Relation to Area, Rainfall and Livestock.

S. No.	District	Area (sq. km)	Ann. Ave. Rainfall (mm)	Livestock[2] (Lacs)	*Gauchar*[1] (ha)
1.	Jodhpur	22850	290	37.89	1,23,000
2.	Nagaur	17718	310	32.27	74,000
3.	Pali	12387	480	27.38	92,000
4.	Jalore	10640	418	18.13	51,000
5.	Barmer	28387	260	41.76	2,13,000
6.	Jaisalmer	38401	140	24.77	89,000
7.	Bikaner	27244	260	25.31	39,000
8.	Churu	16830	320	21.00	48,000
9.	Sikar	7732	440	17.96	48,000
10.	Jhunjhunu	5928	445	12.48	45,000
11.	Sriganganagar	20362	248	14.27	26,000
Total		205941	Range: 140-480	273.16	8,46,000

(Source: 1. Rajasthan Revenue Records; 2. Statistical Abstracts, Rajasthan, 1997)

VI. Present Status of *Gauchars*

Gauchars in general are under great stress. Their size, productivity and support to livestock for fodder is in shambles. A case study of Bikaner district is an eye opener while *gauchars* of other districts present the same scenario. The case of Bikaner district is presented below: -

Gauchars of Bikaner District: A Case Study

	Original Status: (In Bighas)		Present Status: (In Bighas)
1	81 Villages of Bikaner Tehsil Under command area: Lunskaransar *Gauchar + Oran* = 49245	:	Reduced to 15984
2	Sada Nathania *Gauchar*: 27205	:	Now under control of Army
3	Napasar *Gauchar*: 9890	:	Encroached
4	Bheenasar *Gauchar*: 5207	:	Protected by people (Encroachment removed)
5	Jaymal Sar Village: 2416	:	1225
6	Mehrasar Village: 2370	:	Totally lost
7	Kheera Village: 1221	:	190
8	Panpalsar Village: 614	:	Lost
9	Kolayat Tehshil *Gauchars* of 25 villages	:	Lost
10	Chattargarh Village : 2786	:	1200
11	Condition in other villages	:	Decay

VII. *Gauchars* versus Natural Pastures

To sustain population of livestock, especially cows, the natural pastures and fodder species found in different village institutions need special attention. There are several pastures in every district, which need evaluation for flora and fauna and their current capacity to support livestock besides scope of intervention to augment their productivity. In Pali-Digga area of Jaisalmer district, thousands of sq. km. of natural pasture of 'sewan' *Lasiurus sindicus* grass is available. This massive pasture used to feed thousands of cows in drought, as well as, good rain years. Tharparkar, the best cow breed of this region survived because of natural sewan pastures. It is now proven that the sewan grass is perhaps the best fodder species available in Thar. It contains 7-11% of protein with over 8 years of life after harvest. It was evidenced during the worst drought of 1986-87. Over 2 lac quintals of sewan fodder from Jaisalmer was supplied to 10 districts of Rajasthan.

VIII. *Orans* (village forests)

Like *agor* and *gauchar, oran* is yet another important village institution serving Thar for about 8 centuries now. In absence of scientific data, our information is based on 'bahis', oral stories, anecdotes and folklore, war tales, and biographies of warriors, local heros and deities. It has been estimated that most *orans* are about 150-500 years old while some are in existence for 700-800 years. It appears that *orans* came into being concurrently with the founding of a village. Most of the *orans* are dedicated to deities, local heros and martyrs. These *orans* have been named after about 100 such revered people. The most common and popular names are:

Deities:	Bheruji, Hanumanji, Shivji, etc.
Lok 'devtas':	Ramdeoji, Papbuji, Gogaji, etc.
Lok 'devis':	Vankalmate, Karnimate, Jogmaya, Chamunda, etc.
Sects:	Charans, Naths, Shivs, Bheels
Martyrs:	Mamaji ka *oran*

In most *orans,* a temple of a deity or local hero or 'Than' is present to maintain sanctity of the *oran*. Unfortunately, the quality and productivity of *orans* have gone down considerably in the recent past but their sanctity remains. An *oran* is a repository of biological diversity of that area. It supports livestock of one to several villages, besides housing one or more water bodies. Most of the *orans* support 2 or more villages of the area, some 5 or more while we have found a few supporting as many as 30 villages. There are 2-3 *orans*, which support as many as 100 villages around. This speaks for the potentiality of these *orans*. Following are the major *orans* of Jodhpur, Jaisalmer and Barmer districts.

A case study of *orans* of Jalor district is provided here.

Orans of Jalor District: A Case Study

Total villages	706
Villages surveyed	652
Total *Orans*	319 +
Less than 50 ha	257
Process of extinction	54 + mostly encroachments
Orans lost	24%
Good *Orans* sustains	57% livestock
Good *Orans* Support	Average: 2638 people
Without *Orans* Support	Average: 816 People
Oran Villages	Average inhabitants: 894
Poor *Oran* village or no *Oran*	Average inhabitants: 667
Growth of *Prosopis* in *Orans*	Massive
Periphery of *Orans*	Encroached
Water Bodies in *Orans*	187
Most of the *Orans*	Named after Martyrs, Shiv Sect Shakti (Devi), Deities, etc.

Orans are our potential repositories of biodiversity protecting and conserving Thar's germplasm besides supporting livestock, wildlife and avian fauna, serving as water catchments, housing water bodies and acting as shelter belts.

Major *Orans* in Barmer, Jaisalmer and Jodhpur District

S. No.	Name of *Oran* (Tehsil)	Area in Bighas
	I. BARMER	
1.	Dhok Vevetara Mata (Chohatan)	17947.00
2.	Unrod (Sheo)	10463.50
3.	Indroi Mataji (Barmer)	6684.02
4.	Girab (Sheo)	6650.10
5.	Kusip (Siwana)	4377.50
6.	Bothiya-Purohitaan (Barmer)	4165.30
7.	Chuli (Barmer)	2924.19
8.	Nai-Khadali (Guda-Malani)	1711.08
9.	Lakhotali (Barmer)	1537.08
10.	Ramaniya (Siwana)	1394.06
	II. JAISALMER	
11.	Bhadariya ji	42000
12.	Ramdeora (Ramdeoji) (Pokaran)	35171
13.	Eta (Pokaran)	33124
14.	Janara (Pokaran)	13871
15.	Gomat (Pokaran)	11385
16.	Khelana (Pokaran)	8681.12
17.	Parewar (Jaisalmer)	7836.07
18.	Jemla (Pokaran)	7553.16
19.	Khetalai (Pokaran)	6551.12
20.	Bhensala (Pokaran)	5326.02
21.	Beeramdevra (Pokaran)	5216.08
	III. JODHPUR	
22.	Nosar (Osiyan)	9394.1

23.	Bhiyansar (Phalodi)	6819.12
24.	Jakhan (Osiyan)	5980.15
25.	Barsingho-ka-Bara (Phalodi)	55315.19
26.	Tekara (Phalodi)	5468.8
27.	Jalora (Phalodi)	4625.16
28.	Punasar (Osiyan)	4048.5
29.	Dechu (Shergarh)	3562.4
30.	Palina (Phalodi)	3018.16
31.	Tapu (Osiyan)	2545.18
32.	Baori-Barsingha (Phalodi)	2541.19

As found in the case study of *gauchars* in Bikaner district, the case study of orans of Jalor district is no way less alarming wherein 24% of *orans* have been lost and many *orans* are in the process of extinction. Interestingly, *orans* with good ecology support more human and livestock population compared to mutilated *orans*.

IX. *Prosopis* Trees in AGO

Prosopis juliflora, a fast growing exotic species have invaded Thar like a weed. It often arrests the growth of the native flora underneath, bring down biomass turnover. In several *agors*, it has even obstructed flow of water and has contaminated the water.

In several parts, the *Prosopis juliflora* had overpowered all other species in AGO changing the total profile of CPR's and preventing germination of seeds of native plants, succession of native species and obstructing free grazing. Today it is considered enemy number one of our AGO's.

X. Lack of Legal & Administrative Protection of AGO

Very little is written on legal and administrative status of AGO. In absence of legal and administrative protection without punitive provisions for those who encroach, mutilate, deshape or make use of CPR's for individual benefit, the AGO's will continue to decay. The *agors* are frequently tempered and interrupted for variety of physical works by local residents and State-agencies. Hundreds of cases from tehshil to State Revenue Board are being fought concerning AGO's protection and ownership but their legal status remains undefined and unassigned.

In very simple term, *agors* are designated as revenue lands owned by the State through Gram Panchayat which is supposed to monitor, guard and protect it. No data are available on its extent and total area lying under *agors* in Thar.

Gauchar is a revenue land, which finds place in the State Revenue Records. It is designated as grazing land for livestock. This land is under the direct control of the Gram Panchayat, which cannot change or alter its character. However, District Collector (DM) can convert it in public interest for which DM has to provide equal land elsewhere in that village for grazing purposes to be named as *gauchar*.

In the Government land records, *Orans* have not been given their due status, though traditionally it is a revenue land. These are referred as sacred lands dedicated to deities but without any defined legal status. However, the whole village through village panchayat is supposed to work for its upkeep, protection and management. Traditionally it cannot be used for any other purpose or allotted to any one for any purpose except to protect it as a sacred village grove. An *oran* often houses a temple or *Than* where people come for prayers and offerings.

XI. How to Improvise agors, gauchars and orans (AGO)?

Since these CPR's serve critical needs of poor and landless, ecological rehabilitation of the AGO's will serve as a good social object and address the equity and gender issues.

As *agors*, *gauchars* and *orans* are inter-linked and inter-dependent these cannot be handled in isolation. So a comprehensive approach is necessary. We should work to achieve the following objectives:

General

1. To prepare inventory of *agors*, *gauchars* and *orans* taking support from various state agencies and satellite images:
   a. Floristic composition, cover density and carrying capacity.
   b. Average productivity.
   c. Livestock and human pressure and vulnerability.
2. Demand of user groups-stake holders.
3. Comprehensive action plan for social, technical and administrative intervention.
4. Intensive surveys, research and social tools to be developed for AGO's revival. Extensive PRA's as integral part of MGY.
5. To prepare comprehensive document on AGO.
6. AGO be covered under the Panchayati Raj System.
7. Developing nurseries for native herbs, shrubs, grasses and tree species.

Agors

1. To review and revive the rainwater capturing capacity and to initiate physical works like filling pits and removing obstacles to maintain natural flow of rainwater to water bodies.
2. Mapping of existing limits of *agors* and its capacity to serve water bodies.
3. Soil treatment to minimize soil erosion and where it is possible to grow fodder and bush species.

4. Removal of *Prosopis juliflora* trees.
5. To halt encroachments, no physical structures be allowed.

Gauchars

1. Massive plantation of native fodder species, plot by plot, to boost ground cover and reseeding with perennial grasses.
2. To develop efficient, cost-effective workable fencing. Social, rubble, trench, 'Dhorapali', etc. as suggested by concerned village development committees (VDC)/ Gram Panchayats.
3. Separate strategies for *gauchars* of less than 1000 bighas, more than 1000 bighas, 5000 bighas and above but less than 10000 bighas, etc. *Gauchars* above 25000 bighas need special package.
4. New *gauchars* may be developed if agriculture land or any other land is volunteered by individuals, trusts, temples, societies and others with a rider that the fodder and other produce shall be made available free of cost or on subsidized rates as decided by VDC or gram panchayat to the needy.
5. For sustainability and good production in *gauchars,* village had to develop code of conduct and procedures of grazing schedules, to fix animal numbers for grazing, avoid overgrazing, areas to be closed for fodder productions (like fallow) and areas to open for grazing when developed!
6. Make provision for water conservation. Develop new water structures in *gauchars* to harness rainwater for soil moisture and also to water fodder plantation in hot summer. For large *gauchars,* tube-wells may be attempted.
7. To make *gauchars* self-supportive, a certain percentage of land may be used for the production of herbal plants for commercial purposes under the super-

vision of VDC or gram panchayat. The earnings are to be used by village for upkeep, maintenance and infrastructure besides ensuring productivity of soil, etc.

8. During good rain years or when additional fodder production takes place under MGY, the extra fodder be harvested and stored in fodder banks by Gram Panchayat. This fodder can be made available to villagers during lean years or drought at subsidized rates as decided by gram panchayat or VDC (also see section X).
9. Collection and storage of seed material of high yielding perennial grasses and shrubs for use in *gauchar* development.
10. To record peoples perspective on *gauchars*.

Orans

1. Efforts are to be made to revive *orans* linking them with the cultural fabric of desert society.
2. To provide them their due legal status.
3. To study and research on soil profile, water quality, water retention capacity availability of animals and plant taxa, their inter-relationship, interdependence and diversity dynamics to optimize *oran's* productivity.
4. Revival of lost cycles of trees where plant successions have collapsed.
5. Planting of native plant species including herbal plants.
6. Revival with careful scientific insight, inputs and intervention after seeking approval from Gram Panchayat or VDC.
7. To ensure and encourage social fencing.
8. No mining leases in and around *orans*.

9. Special care, upkeep and extension of water catchments and water bodies located in *orans*.
10. Total removal of *Prosopis juliflora* trees.

Literature consulted

Adams, A. 1899. (reprint 1900). *The Western Rajputana States*. xi + 455 pp. Jodhpur, (Vintage Books).

AFRI, 2002. Proceedings of Workshop: On Development of Suitable Strategy for Rehabilitation of Orans and Gauchars in Rajasthan. Director, Arid Forest Research Institute, 125 pp. Jodhpur.

Anonymous. 1998. *Thar Desert Biosphere Reserve*. Ministry of Environment and Forests, Govt. of India. Project Document No. 7. 1-100 pp.

Berker, F.; Feeney, D.; Mc Cay, B.J. and Acheson, J.M. 1989. The Benefits of the Commons. *Nature*, 340: 91-92.

Bhandari, M.M. 1970. Famine foods of Indian desert. *Eco. Bot.*, 28: 73-81.

Bhandari, M.M. 1990. *Flora of the Indian Desert*_(2nd Ed.). MPS Repros, Jodhpur.

Bhandari, M.M. 1995. Biodiversity of the Indian Desert. In Taxanomy and Biodiversity (ed. A.K. Pandey), pp. 29-43. Delhi (CBS Publishers).

Bharara, L.P. 1999. Man in the Desert. Drought, Desertification and Indigenous Knowledge for Sustainable Development. Scientific Publishers (India), 394 pp.

Chakravarty, A.K.; Ratan, R. and Singh, K.C. 1970. Grazing studies in the arid and semi-arid zones of Rajasthan: Utilization of vegetation cover, grazing behaviour of sheep, and seasonal variation of crude protein content of plants in different pastures. *Ann. Arid Zone*, 9: 10-18.

Direcor, Z.S.I. (ed.) 1991. Animal Resources of India. Protozoa to Mammalia: State of the Art. Z.S.I. Calcutta.

Erskine, K.D. 1908 a. *Rajputana Gazetteers. The Western Rajputana States, Residency & Bikaner*. 104 pp. Gurgaon (Vintage Books). (reprinted 1992).

Erskine, K.D. 1908 b. *Rajputana Gazetteers. The Western Rajputana States, Residency & Bikaner*. xii+400 pp. Guragaon. (Vintage Books). (reprinted 1992).

Feeny, D.; Berkes, F.; Mc Cay, B.J. and Acheson, J.M. 1990. The tragedy of the commons: Twenty-two years later. *Human Ecology*, 18: 1-19.

Gadgil, M. and Guha, R. 1992. *This Fissured Land: An Ecological History of India*. Berkeley, University of California Press.

Gadgil, M. and Vartak, V.D. 1975a. Sacred groves of India: A plea for continued conservation. *J. Bomb. nat. Hist. Soc.*, 72 (2): 312-20.

Gadgil, M. and Vartak, V.D. 1975b. The Society of groves of Western Ghats in India. *Economic Botany*, 30: 152-60.

Gadgil, M. and Vartak, V.D. 1981. The Sacred Groves of Maharashtra: An Inventory. In Glimpses of Indian Ethnobotany. (ed. S.K. Jain). Delhi (Oxford and IBH Publ. Co.).

Gadgil, M. and Vartak, V.D. 1994. *The Sacred Uses of Nature. Social Ecology*. Delhi, Oxford Uni. Press.

Ghosh, A.K.; Baqri, Q.H. and Prakash, I. 1996. *Faunal Diversity in the Thar Desert: Gaps in Research*. xi+410 pp. Jodhpur (Scientific Publishers).

Gold, A.G. and Gujar, B.R. 1989. Of gods, trees and boundaries: Divine conservation in Rajasthan. *Asian Folklore Studies*, 48: 211-229.

Hardin, G. 1968. The tragedy of the commons. *Science*, 162: 1243-48.

Hooja, Rakesh and Joshi, R. 1999. (ed.). *Desert, Drought and Development: Studies in Resource Management and Sustainability*. Rajasthan Studies: Rawat Publishers, Jaipur, 386 pp.

Jain, Bhuvnesh. 2002. Personal Communication.

Jodha, N.S. 1985. Population growth and decline of common property resources in Rajasthan, India. *Pop. Devel. Rev.*, 11: 247-64.

Jodha, N.S. 1986. Common Property resources and rural poor in dry regions of India. *Eco. Pol. Weekly*, 21: (26).

Malhotra, S.P. 1988. *Man and the Desert. In Desert Ecology* (ed. I. Prakash,), pp. 37-64. Jodhpur (Scientific Publishers).

Maloo, Kamla. 1987. *The History of Famines in Rajputana (1858-1900 A.D.)*. Himanshu Publications, New Delhi. 296 pp.

Meena, V.P. 2000. *Desert Ecology*. Pointer Publishers, Jaipur, 195 pp.

Mohnot, S.M. and Bhandari, M.M. (ed.). 1986. *Environmental Degradation in Western Rajasthan*. Jodhpur (Jodhpur Univ. Press).

O'brien, A.G. 1996. The Ancient Chronology of Thar. The Bhattika, Laukika and Sindh Eras. Oxford University Press, Delhi. 211 pp.

Panwar, L.K. 1986. The Battle of Survival in Jaisalmer: Sevan Story. In Environmental Degradation in Western Rajasthan (ed. S.M. Mohnot and M.M. Bhandari), pp. 113-118. Jodhpur, Jodhpur Univ. Press.

Panwar, L.K. 1998. Ecological and Wildlife Studies in Indian Desert to Boost Tourism. Ph.D. Thesis of Rajasthan University, Jodhpur. 207 pp.

Panwar, L.K. 2001. *Bhartiya Registhan Mai Paryavaraniya Paryatan*. Rajasthani Granthagan, Jodhpur, 218 pp.

Parihar, G.R. 1997. *Studies on the Biodiversity of Some Selected Areas in the Desert*. Ph.D. Thesis of JNV University, Jodhpur. 148 pp.

Patwa, Subhu. 1989. *Paryavaran ki Sanskriti*. Vaghdevi Prakashan, Bikaner, 123 pp.

Rao, J.V. Ramana. 1994. *The Living Desert*. Defense Scientific Information and Documentation Centre (DESIDOC), 129 pp.

Robbins, Paul. 1996. Negotiating Ecology: Institutional and Environmental Change in Rajasthan, India. Ph.D. Thesis, Clark University, Worcester, MA, USA.

Roonwal, M.L. 1977 (ed.). *The Natural Resources of Rajasthan*. Vol. 1 & 2. The University of Jodhpur, 1211 pp.

Sen, A.K. and Kar, Amal 1993 (ed.). *Desertification and its Control in the Thar, Sahara and Sahel Regions*. Scientific Publishers, Jodhpur.

Sen, A.K. and Kar, Amal 1995 (ed.). *Land Degradation and Desertification in Asia and the Pacific Region*. V+365 pp. Jodhpur. (Scientific Publishers).

Shankarnarayan, K.A. 1984 (ed.) *Agro-Forestry in Arid and Semi-Arid Zones*. CAZRI, 295 pp.

Sharma, N.K; Singh, R.P.; Yadav, M.S. and Singh, K.C. 1999. *Forage Production in Dryland of Arid and Semi-Arid Regions*. Scientific Publishers (India), 152 pp.

Singh V. and Pandey, R.P. 1998. *Ethnobotany of Rajasthan, India*. Scientific Publishers (India), Jodhpur, 367 pp.

Singh, L.N.; Singh, A. and Singh, J. 1998. *Forage Grasses and Legumes*. Scientific Publishers (India), Jodhpur, 133 pp.

Singh, Shekar; Sastry, A.R.K.; Mehta, Raman and Uppal, Vishasish 2000. (ed.). *Setting Biodiversity Conservation Priorities for India. Vol. 1*. World Wide Fund for Nature, India, 337 pp.

Singhvi, A.K. and Kar, Amal 1992. (ed.). *Thar Desert in Rajasthan. Land, Man and Environment*. Geological Society of India, 191 pp.

Statistical Abstract: Rajasthan. 1997. *Directorate of Economics and Statistics, Rajasthan*, Jaipur. 377 pp.

EXTENSION OF ECOLOGICAL PRINCIPLES IN REHABILITATION OF LIGNITE MINED WASTELANDS

Suresh Kumar and Parveen Kumar
Central Arid Zone Research Institute, Jodhpur

ABSTRACT

Rehabilitation of mined land is guided by ecological practicability and social value judgement and hence each site needs separate strategic planning to have natural, near-natural or semi natural ecosystem. Land shaping and characterization of spoil surface decide about the soil additives for profile modification. Based on soil and climatic conditions, suitable plant species could be selected for plantation based on a number of criteria which are discussed in detail. Saplings of suitable species are prepared well in advance and planted on the mine site in March. Details of plantation, irrigation, weeding and after care are presented in this paper. Survival of species planted at lignite mine spoil was 82.82%. Maximum growth was observed in *Parkinsonia aculeata* followed by other species. It has been concluded that by extending ecological principles, success in rehabilitation could be achieved in the desired time frame. Role of local people and NGOs for participatory mine rehabilitation has been highlighted as a major future requirement of these programs.

1. Introduction

Rehabilitation is returning land to a form and productivity in conformity with reclamation and landuse plan for the site. This includes isolation of disturbance impacts to the site, limiting contribution to environmental deterioration and making site amenable with surrounding aesthetic values. Rehabilitation is different from 'restoration' and 're-vegetation'. 'Restoration' is re-establishment of the same ecosystem as in the pre-disturbance landscapes. Re-vegetation is establishment of vegetative cover following land disturbance which is a component of overall reclamation process. 'Reclamation' is a process of reconverting disturbed lands to their former uses or other productive uses as per defined post mine land use. Reclamation aims to achieve cost efficient and biologically effective modification of the site to limit offsite effects with long-term stability. Though rehabilitation ecology has wider scope covering all types of disturbances and disasters i.e. floods, earthquakes, fire, frosts, browsing, grazing and cutting, we will concentrate mainly on mining induced disturbances.

2. Ecological Principles in Rehabilitation

In order to apply ecological principles for success in rehabilitation, it is a pre-requisite to understand impacts of mining on landscape and its consequent degradation.

2.1 Impacts of Mining

Mining is excavation of mineral ores, its concentration and processing of final products. During the process, particularly in open cast mining, huge volume of solid waste (overburden/tailings) is generated. Its impacts are both immediate and long term. Immediate impacts include creation of two types of landscapes: the overburden dumps due to stock piling of mineral waste on fertile land. Secondly, open cast mining ruptures the vast stretches of landscapes to end up in large open pits. When these are filled again with mine waste to make up to the surface, these are called 'backfills'. In creating both dumps and backfills, the rock-soil-plant establishment is disturbed, resulting in ecological chaos (Harthill and Mckell, 1979).

Heavy machinery used in mining adversely affects the overburden due to its compaction. These equipments dump the spoil and roll over it to smoothen its central part and the peripheral area remains relatively loose. The central axis of the pit is markedly packed while packing is less on out slopes where materials are pushed or slide down on slopes (Power *et al.*, 1978). This causes uneven subsidence, undulating surface, differences in surface drainage and finally slope wash that adversely affects the adjoining land.

Soil disturbance adversely affects many beneficial soil characteristics, which require hundred to thousands of years to develop and attain steady state of maximum soil productivity. The pedological and biological soil loss due to mining results in a surface that is completely devoid of fertile soil, flora and fauna, to remain completely barren for all times. These immediate impacts trigger long term effects of declining productivity in adjoining lands, disrupting hydrological processes and enhancing erosional hazards, culminating in water and soil pollution and occupational hazards affecting ultimately the human health. Thus, these drastically disturbed ecosystems are extremely vulnerable to future anthropogenic changes at both, local and global level (Peters, 1985).

2.2 Extent of mining in India and Rajasthan

In India, some 8,00,000 ha area is affected by mining which not only brings out 623 m tons of minerals but also 3 times more, 1870 m tones of overburden posing a grave environmental threat. 65 major and 22 minor minerals are produced. Rajasthan produces 42 major and 22 minor minerals. Wollastonite and jespar are only in Rajasthan, which has 97% of Zinc, 95% of Gypsum, 96% of fluorite, 70% of sand stone and limestone, 89% of asbestos, 90% of marble, 70% of calcite and 80% of soapstone deposits of India. Some 15000 to 20000 crores of minerals are produced annually in Rajasthan of which government earns a royalty of 400 crores. It provides employment next only to agriculture, 30 lakh directly and 3 million indirectly.

Thus mining has been causing environmental concern by creating drastically disturbed lands on one hand and the livelihood to millions of poor on the other hand. The huge revenue it provides, make it a social compulsion and political propriety to its continuance despite judiciary's strictures. In such a scenario, it becomes essential to find out techniques of rapid rehabilitation modules of mining of different minerals so as to minimize deleterious impacts of mining. Concepts and application towards this goal are described below.

2.3 The Ecological Principles

Rehabilitation of mined land is guided by ecological practicability and social value judgement and hence each site needs separate strategic planning to have natural, near-natural or semi natural ecosystem. Thus, the process of rehabilitation should be allowed/made to proceed towards multiple end points. These multiple end points will decide prioritization of different activities of rehabilitation. But the first step is to protect the disturbed habitats and communities to prevent further degradation and loss of extent genes. Thereafter, attempts are made to accelerate the re-vegetation process for increasing biodiversity and stabilizing nutrient cycling. Again for re-vegetation, primary action is to control erosion through plant cover in the short term and development of a self sustaining community through colonization of native species in the long term (Mitchell, 1987). Success will depend much on augmentation of biological activity of the soil surface horizon in the long term (Arnold *et al.*, 1984). Restoration of physical structure, chemical composition and biological functioning of mine spoils by restructuring the vegetation is the final strategy. There are two different path ways of rehabilitation strategy:

1. To allow minimum agronomic inputs thus permitting natural biological succession to reign.

2. To pump in maximum agronomic inputs i.e. tillage, fertilizers, liming, etc. to achieve rapid plant cover.

Blending of the above two approaches is quite often the most desirable way (McKendrick, 1980). Finally, the rehabilitation plan must determine the:

- uses, the site can tolerate
- relative importance of uses
- possibilities for improvement
- economics of various management practices
- physiognomic and ecological consequences of the alternatives
- rates of change
- capacities for manipulating the environmental systems to particular end points
- required aftercare.

3. Extension of Ecological principles in field level rehabilitation

3.1 Characterization of spoil

The extension of ecological principles in rehabilitation of mine spoil begins with charactering the overburden muck. The physical and chemical analysis of muck material following standard analytical tools will reveal the limitations and possibilities of its rehabilitation. This will help in deciding three things:

(1) The type of soil amendments to make it a growing media.

(2) The type of soil and water conservation measures required and

(3) The type of species suitable for rehabilitation of lignite spoils.

Analysis of lignite mine backfill revealed that compared to normal soil, the backfill had high silt and clay, very high sodium and high electrical conductivity (Praveen Kumar *et al.*, 2005). Since gypsum addition would require water in huge quantities for neutralizing high sodium, this option was not followed. Instead, simple layering with original top soil emerged as cheapest alternative.

3.2 Landscaping and restructuring the mine spoil surface

For reclamation, mining muck heap/hill or backfill need to be restructured first. Mining muck heaps normally vary from place to place and type of mining in respect of height, slope, width, thickness, size of aggregates and voids besides their chemical constitution. In case of systematic dumping the height may range from 10 to 40 m.

Mining dumps being devoid of topsoil can hardly support any vegetation. Surface modification of the spoil through chemical means is very costly. Hence, cheapest way is to cover these spoils with topsoil from neighbouring areas. The thickness of such topsoil varies from 5-60 cm. Other soil conservation measures like land leveling, surface compaction, stabilizing the piles, terracing, contour furrowing and bunding can be adopted there after. This also fills the voids to make a perfect rooting medium. Further, seed bank in soil will help regenerate the natural weed flora. Their short duration growth will give anchorage to soil with their shallow root system. Almost all the muck heaps are deficient in nitrogen, phosphorus and potash. Hence, for successful establishment and reasonable growth of transplanted material farmyard manure (FYM) together with chemical fertilizer, etc. may be used.

3.3 Water harvesting and excavation of planting pits

The operation is started well in advance before rainy season. The surface is reshaped into terraces and slopes for providing catchments for rainwater harvesting and terrace for transplanting. The surface of the mined wastelands can be reshaped to micro catchments, half moon terraces, ridge & furrows and inward sloping bench terraces. Planting pits (60 cm^3) excavated at 3 m spacing in rows 9 m apart, are filled up with necessary growing medium, consisting of mixture of fine sand /farm soil and farmyard manure in 2:1 ratio.

3.4 Selection of Plant Species

Based on bio-climatic conditions, soil type and desirable post mine use, the species selected should have the following characteristics:

1. It can be propagated easily.
2. It may be established successfully.
3. It should be compatible with other species.
4. It should be able to provide quick vegetative cover of pioneer species.
5. It should be able to afford revival and growth of other desirable species, can be called also as nurse crops.
6. It is a succession facilitator species.
7. It should have desired rooting habit.
8. It should serve as feed for wildlife and/or livestock.
9. It should have tolerance to utilization.
10. It should possess strategic adaptations in the phonological cycle and morphology so as to be able to evade/adapt critical periods of extremes of temperatures, drought and salinity.
11. It should have competitive edge over invaders.
12. It should be reasonably resistant to pest and microbial infections.
13. It should have minimum or no requirement of externally added fertilizers.
14. It should have rapid seedling growth and seedling drought/frost tolerance.

15. It should have overall wide ecological amplitude in its life cycle strategy.

Using a combination of above criteria, multipurpose native species of trees, shrubs, and perennial grasses for arid and semi-arid areas should be planted. Some of the exotic species, whose performance has been well established in arid tract, may also be included for the rehabilitation of mine spoil (Table 1). Leguminous trees and shrubs should be preferred.

Since majority of lands prior to mining in arid tract are grazing ground or rainfed agriculture, post mining land development into agroforestry and/or silvopasture system is the preferred choice.

Table: 1. Suitable plant species for revegetating lignite mine spoils

Type of Vegetation	Plant Species
Trees	*Acacia senegal, Acacia tortilis, Azadirachta indica, Leucaena leucocephala, Parkinsonia aculeata, Prosopis cineraria, Salvadora oleoides, Tecomella undulata.*
Shrubs	*Cassia angustifolia, Circidium floridum, Dichrostachys nutans, Tamarix aphylla, Ziziphus rotundifolia, Acacia nubica, Colophospermum mopane.*
Grasses	*Cenchrus setigerus, Cenchrus ciliaris, Lasiurus sindicus.*

3.5 Planting on spoils

Healthy seedlings of carefully chosen species are prepared in nurseries in poly bags in advance. Four months to an year old seedlings are kept ready for planting at the mining site with suitable soil working for increasing water availability to seedling. Trees and shrubs could be planted at 3 to 5 m in a row which could be 6 to 9 m apart. The interspaces can be planted with annuals, grasses, legumes or crops.

Weeding, soil working and watering (15 l/plant) in first three years is done as per requirement. No watering is done from June end to October.

3.6 Use of mulches and stabilizer

Mulches are required to reduce evaporation of water from soil surface in the pit while stabilizers are placed at the bottom of the pit or mixed with potting media to prevent water loss due to deep drainage. Commonly used mulches are hay straw, crop stubble and even gravel. These can not be utilized, however, on sloping land. Commonly used stabilizer, a polymer, *jalshakti*, when mixed with soil in planting pit will help in retaining more moisture for longer period and finally boost the plant growth. Bentonite placed as subsurface barrier also prevents deep drainage. Pond silt, farm yard manure and local soil (not spoils of mine waste) are also added to make the spoil as a better growing medium.

3.7 Post plantation care

Beside watering at the time of plantation and at regular intervals, we need to do weeding and hoeing to ensure its proper growth. The young sapling's leaves are much liked by squirrels, rabbits, peacocks, chinkaras and stray domestic livestock. A protection of thorny sticks as canopy over saplings will act as a deterrent to animals. This canopy should be removed when plant attains 50-70 cm height. Lower branches should be cut to train sapling into a single stem tree.

3.8 Survival

Survival of different plant species varied in different treatments, the overall average being 82.28% in backfill plantation (Fig. 1). The minimum survival was 50% in *Tamarix aphylla*. Species showing over 80% survival included *Salvadora oleoides, Acacia nubica, Tecomella undulata, Prosopis cineraria, Azadirachta indica, Colophospermum mopane, Acacia senegal, Parkinsonia aculeata, Cercidium floridum* and *Leucaena leucocephala.*

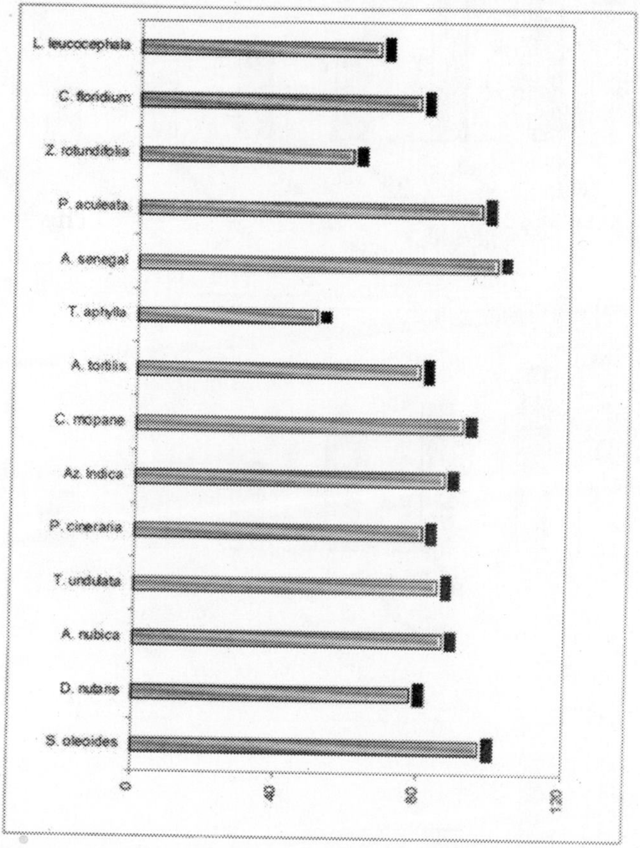

Fig. 1. Overall survival of different species at backfill

3.9 Growth

Height, collar diameter and canopy cover increase per year was maximum in *Parkinsonia aculeata* (Figs. 2, 3, 4) followed by *Acacia tortilis, Salvadora oleoides, Dichrostachys nutans, Acacia senegal, Tecomella undulata* and finally *Prosopis cineraria* respectively. The saplings planted in March grew better than those grown in July – August.

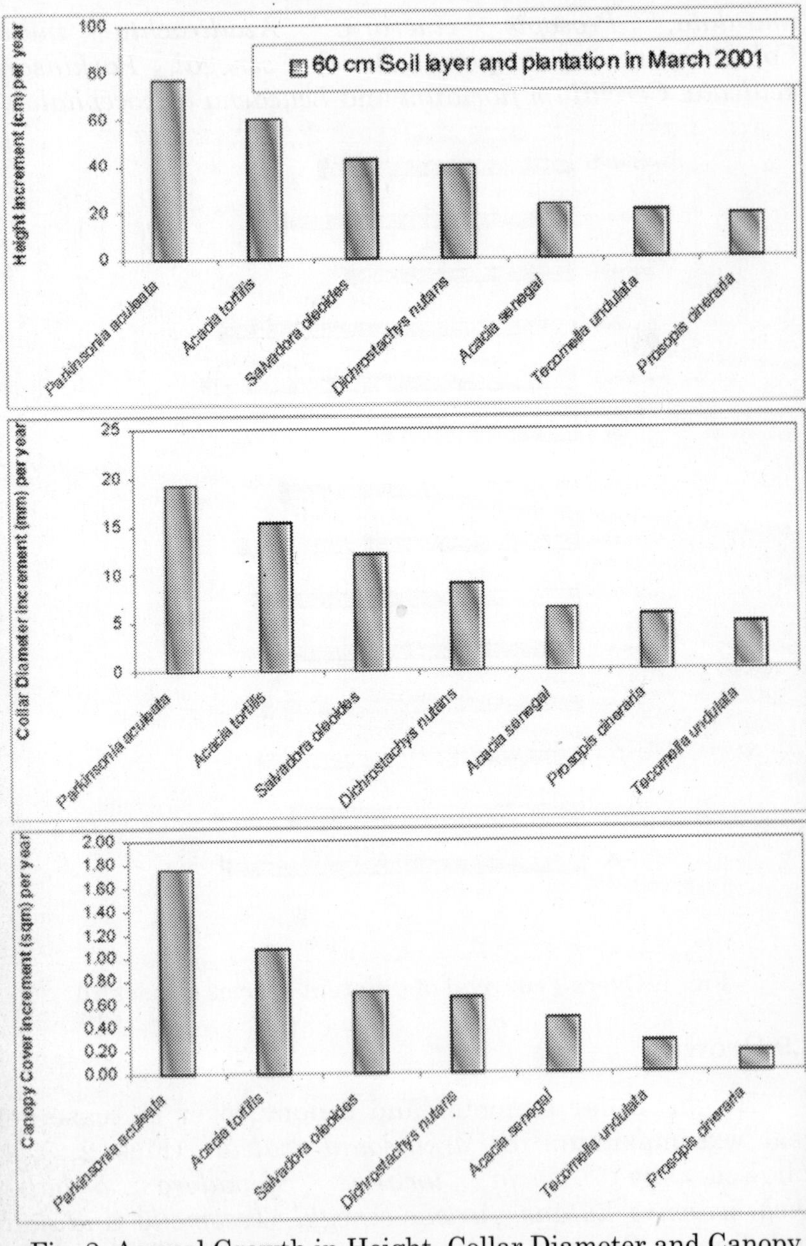

Fig. 2: Annual Growth in Height, Collar Diameter and Canopy Cover of different species planted in March 2001 on 60 cm layer of soil and murram at Backfill, Giral.

Fig. 3: Heaps of overburden at lignite Backfill in foreground before rehabilitation with high dumps behind at Giral, Barmer

Fig. 4: Nine months after planting *Parkinsonia aculeata* at back fill. *Cenchrus ciliaris* growing in the inter row space after rehabilitation

3.10 Conclusion

It can be concluded that lignite mine spoil backfill can be rehabilitated by layering 60 cm with local soil, suitable modification of pit soil and plantation of ecologically matching species in the month of March. The success of rehabilitation program was because of following ecological principles applied at Giral

- Site - species suitability
- Soil & water conservation
- Soil seed bank
- Soil fertility / Toxicity
- Succession
- Density – dominance (vegetation ecology)
- Population Ecology
- Equilibrium and climax
- Ecological economics.

Future Strategies

Rehabilitation of lands disturbed by mining has become obligatory for both; the state owned corporations as well as private companies in view of environmental concern and recent stand taken by the Judiciary. As newer and more areas will be brought under mining, the need for rehabilitation of these lands will become more obvious. Realizing this, the protocols of rehabilitation perfected by CAZRI will prove to be extremely useful. Refinements for fine-tuning these protocols to suit a particular edapho-climatic situation will be therefore, our future priority in the larger interest of our economy and environment. In order to make rehabilitation programme cost effective and people friendly, participation of local people in providing social fencing, as also generating employment for them should receive focused attention of our policy makers, miners and the NGOs.

References

Arnold, P.W., Gildon, A. and Rimmer, D.L. 1984. The use of soil in reclamation of coal mining wastes. pp. 26.1-26.10. In

Proceedings of symposium on reclamation, treatment and utilization of coal mining wastes. Durham, England.

Harthill, M. and McKell, C.M. 1979. Ecological stability - is this a realistic goal for arid land rehabilitation? pp. 557-567 In: M.K. Wali (ed.) *Ecology and Coal Resource Development,* Vol 2 Pergamon Press, New York.

Kumar, Praveen; Kumar, Suresh; Sharma, K.D.; Chaudhary, A. and Gehlot, D. 2005. Lignite mine spoil characterisation and approaches for its rehabilitation. *Arid Land Research and Rehabilitation*. 19:47-60.

McKendrick, J.D. 1980. Mine reclamation in portions of West Germany, Union of Soviet Socialistic Republics and Alaska. *Agroborealis* 12:11-14

Mitchell, W.W. 1987. Re-vegetation research on coal mine over-burden materials in interior to south eastern Alaska. *Agricultural and Forestry Experiment Station Bulletin* 79:72-86.

Peters, R.L. 1985. Global Climate Change: A challenge for restoration ecology. *Restoration and Management Notes*. 3: 62-67.

Power, J.F., Sandoval, F.M. and Ries, R.E. 1978. Restoration of productivity of disturbed land in the Northern Great Plains. pp 33-49 In R.A. Wright (Ed) *The reclamation of Disturbed Arid lands*. University of North Mexico Press, Albuquerque.

FORESTRY RESEARCH EXTENSION IN RAJASTHAN: PRESENT STATUS, POTENTIAL, EXPECTATIONS, CONSTRAINTS, PROBLEMS AND STRATEGY FOR FUTURE APPROACH

K. Kala RFS*

Rajasthan Forestry & Bio-diversity Project,
Forest Department, Rajasthan, Jaipur.
*Deputy Conservator of Forest, Communication,
Extension & Training, Raj. Forestry and Bio-diversity Project,
Jaipur (Rajasthan.)

ABSTRACT

Forest play a vital role in the economy of the state in fulfilling the essential needs for fuel, fodder and small timber of local communities and also a very crucial role in improving the socio-economic conditions of the rural people, besides ameliorating environmental conditions. About 70% population is living in rural area and engaged mostly in agriculture and livestock rearing. Due to uncertain and erratic rainfall, people rely more on pastoral economy than agriculture. Rajasthan is lagging far behind with recorded forest area equal to 4.19% of the country's Forest area. For the state of Rajasthan where arid and semi-arid geo-climatic conditions prevail over a large tract, the rural economy is essentially pastoral; and a sizeable livestock population creates heavy biotic pressure on forest areas, the sustainable development of forest resources assumes a special significance. Due to harsh climatic and difficult edaphic factors greening is quite a challenging task in the state. Forest development in such circumstances requires a well thought out strategy for each eco-system.

The department has planned for the short, medium and long-term perspective of forestry development in the state. The forest cover is proposed to be increased on fallow lands, degraded private lands, community lands and waste lands besides expanding the forest cover on the recorded degraded forest lands and on arable land through Farm Forestry. In order to achieve the policy goals it is necessary to have research studies to evolve suitable models to reclaim/ develop the waste lands situated in as well as out side the forests lying in the form of water logged areas, alkaline and saline soils, denuded and degraded forest lands, Gauchars, Orans, Panchayat and private waste lands. In order to get tree cover enhanced, it will be needed to evolve suitable Agro forestry, Agri-silvi-pastoral models of different combinations of tree species and agriculture crops for different agro climatic zones. Research studies are also needed in the field of protection as well as production forestry i.e. protecting of natural stand from insect and micro organic pests, rodents pests etc. and to take maximum yield from production of bio-mass etc. For dissemination of research results, the research findings will be properly documented in journals, newsletters and pamphlets. Developed technology will be transferred to the field for efficient use by the users. Looking in to the need of the hour, we will have to have a strong and detailed future strategy to act firmly to address the challenges. The present paper deals with the various aspects of Forestry Research Extension in Rajasthan inter-alia present status, potentials expectations, constraints and problems and strategy for future approach.

Introduction

Rajasthan, the largest state in the country, covers 34.22 million ha, which is 10.41% of the country's geographical area comprises of arid and semi-arid zones.

It lies between 23°30" and 30°11" North latitude and 69°29" and 78°17" East longitudes. The Aravallis, one of the oldest mountain systems, covering 30% of the state, divides

Rajasthan into two unequal parts. The precipitation in this hill range is not high and soil cover is insufficient. The area is embedded with ravines and depressions. A vast expanse of the arid and semi-arid tract- the Thar Desert, lies in the west of the Aravallis and covers about 20.87 million hectare area alone in the state.

The climate is accordingly varied. Hyperthermic conditions prevail in the entire state. The mercury touches 53°C at some of the places during summer and drops below freezing point during winter. The rainfall pattern in the state is very erratic. The annual rainfall received is as low as 150 mm in the extreme arid zones and as high as 1000 mm in the southeastern part of the state. The ravines and plains are subjected to recurring floods and droughts.

In-spite of inadequate forest area and forest cover, the tropical dry deciduous and thorny forests support rich variety of flora and fauna including many globally endangered mammals and avi-fauna like Black-bucks, Great Indian Bustard, Bengal and Lesser Floricans etc. in the state.

Physiography and Major Ecosystems

(1) Physiographic Regions

The physiography of Rajasthan is the product of long years of erosional and depositional processes. State's area can be identified and classified into major physiographic regions as below:

The Western Desert.

- Aravalli Hills.
- The Eastern Plains.
- The Banas Plains.
- The Chappan Plains.
- The Southeastern Plateau.

(2) Ecosystems

Depending upon the climatic conditions and forest vegetation, the state is divided into four major fragile ecosystems as follows:

S.N.	Ecosystem		Sub Type
1.	Desert Ecosystem	i.	Canal command area.
		ii.	Non command area.
		iii.	The Luni basin.
2.	Aravalli Hill Ecosystem	i	Northern Aravalli region.
		ii	Central Aravalli region.
		iii	Southern Aravalli region.
3.	Eastern Plains Ecosystem	i	Banas basin.
		ii	Mahi basin.
		iii	Banganga basin.
		iv	Sahibi basin.
		v	Gambhir basin.
		vi	Varah/barah basin.
4.	Hadoti Plateau and Ravine Ecosystem	i	Chambal basin.
		ii	Dang area.

Present status of forests

(1) Forest Types

The forests of Rajasthan are basically of five types spread unequally in northern, southern, eastern, western and south-eastern parts. The state has teak forests, which is northern most limit of teak zone in India. The forests are mostly in edapho-climatic climax stage of succession. Area by forest types is as under:

Table 2. Area by Forest Types

S.N.	Type	Forest area (SQ KM)	%
i.	Dhok (*Anogeissus pendula*) forests	17868.40	55.00
ii.	Mixed Miscellaneous forests	6595.06	20.30
iii.	Dry teak forests	2306.65	7.10
iv.	Salar forests	1656.89	5.10
v.	Tropical thorn forests	1656.89	5.10
vi.	Khair forests	974.64	3.00
vii.	Bamboo forests	812.20	2.50
viii.	Palas forests	487.32	1.50
ix.	Tropical semi-ever green forests	129.95	0.40
	Total	32488.00	100

(2) Forest cover

Due to the adverse geo-climatic conditions of the state, forest cover in the state is minimal, which is only 4.84% (1.58 million hectare) of the state's geographical area. Besides, there is an unequal distribution of forest area in the state. Most of the forests are concentrated in the southern and southeastern parts of the state, with hardly any forests in the western parts. These are dry deciduous in hilly terrain and Tropical Thorn Forests in arid and semi-arid regions of the state.

According to FSI (2003), total forest cover of the state, based on density, can be classified into four categories:

- Very Dense Forest (crown density more than 70%) is only 0.0014 million ha.
- Moderately Dense Forest (crown density between 40% and 70%) is only 0.448 million ha
- Open forest (crown density between 10% and 40%) is 1.133 million ha.

- Scrub and barren forest (crown density less than 10%) is 0.456 million ha.

Per capita forest cover in the state is 0.0281 ha, far below the national average of 0.066 ha. The total growing stock of the state forests is 31.960 million cu. m. (FSI, 2003). Thus, average volume of the state works out to 9.84 cu m/ha as against 61.72 cu m/ha for the country.

The Forest Department, Government of Rajasthan has planned for the short, medium and long-term perspective of forestry development in the state. The "State Forestry Action Programme (1996-2016)" proposes to increase the forest cover from the 3.83% (1996) to 20% during the period of twenty years. The forest cover is proposed to be increased on fallow lands, degraded private lands, community lands and waste lands besides expanding the forest cover on the recorded degraded forest lands and on arable land through Farm Forestry.

"Scientific studies have indicated that present feed and fodder availability is deficient by about 36% even for maintenance of current livestock population in the arid and semi arid regions. However, this deficiency rises to about 80% if we want to maintain the livestock population at their potential level of production. Therefore, there is an urgent need for taking necessary steps for halting this degradation process and reviving the potential productivity of the grasslands" (Kapoor, 2005).

Besides these, the right holders and tribals locally collect NTFP, and the demand for NTFP has not been assessed commodity-wise. Tendu leaf is the major source of revenue and of employment. Approximately, 5 million tones of standard bamboos are extracted directly and indirectly from forests (Ghose, 1996).

It has now been recognized that most of the forest areas outside the PA network are largely degraded and, unless confidence of the local people is gained and their support in protection and management (Joint Forest Management) enlisted, it is absolutely impossible to regenerate degraded forests while simultaneously meeting the demands of the local people.

Forestry Research and Extension in Rajasthan

Forestry Research and Extension are like two bullocks of a cart. Without proper extension activities, transfer of research findings (developed technologies) remains without reaching to the users and remains of no use. Like-wise without research, extension of what! So there is need to run both simultaneously, with well thought-out plan or policy, then only the desired results are possible.

(1) Need Analysis

In India more than 175 million ha of arable land is facing degradation of different kinds. Such an alarming scenario of forest and land degradation has resulted in acute shortage of fuel-wood, fodder, timber and other non-wood-forest-produce (NTFP), which supports rural economy to the greatest extent. The decline in output of forest products has generated socio-economic conflicts of very serious nature.

Sustainable development of natural resources is a major concern today. As envisaged in the country's National Forest Policy, 1988, mega bio-diversity rich country like India needs to have minimum of 33% of it's geographical area under forest and tree cover as against present 23.68% to fulfill the ever increasing demand for fuel, fodder, timber and NTFP, to foster agriculture and pastoral economy, besides ameliorating environmental conditions and supporting it's faunal and floral diversity as well.

Rajasthan is lagging far behind with recorded forest area equal to 4.19% of the country's forest area. For the state of Rajasthan where arid and semi-arid geo-climatic conditions prevail over a large tract, the rural economy is essentially pastoral; and a sizeable livestock population creates heavy biotic pressure on forest areas, the sustainable development of forest resources assumes a special significance.

Due to harsh climatic and difficult edaphic factors, greening is quite a challenging task in the state. Forest development in such circumstances requires a well thought out strategy for each eco-system.

With the ever-increasing human and livestock population, pressure on the forest has increased manifold, resulting in their decimation to a level of no recovery. The enormous pressure has resulted in shrinkage of resources' base. This trend needs to be halted by systematically harnessing science and technological inputs ensuring inter-departmental co-ordination and involving beneficiary communities through a mechanism of usufructs sharing, *i.e.*, Joint Forest Management.

The existing forest resources are grossly inadequate to meet the forest biomass based needs of the local communities and the state. It is essential to undertake appropriate strategies for conservation, management and sustainable development of the forest to bring them to an optimum level of productivity and expand forest cover outside the traditional forest boundaries to achieve the policy goals of having 20% of geographical area under forest or tree cover.

The SFAP aims at covering 5.14 million ha area during 20 years period enhancing potential productivity from level of 9.84 cubic meter per ha (1996) to 15.00 cubic meter/ha, thereby increasing the growing stock from 12.89 million cu m (1996) to 82.22 million cu m. The current growing stock is 31.960 million cu m (FSI, 2003).

In order to achieve the policy goals, the emphasis has to be given on the Participatory Forest Management, Micro-planning or Preparation of Community Action Plan, Watershed/Cluster Approach of afforestation and management, HRD/ Capacity Building, Institutional Strengthening – technically qualified and professionally competent manpower & complete restructuring of the present organization,

Forestry Research, Monitoring and Evaluation, Forestry Extension, Harmonization of cross sectoral issues. It is necessary to have, inter-alia, research studies to evolve suitable models to reclaim/ develop the waste lands situated inside as well as, out side the forests lying in the form of water logged areas, alkaline and saline soils, denuded and degraded forest lands, gauchars, orans, panchayat and private waste lands. In order to get tree cover enhanced, it is needed to evolve suitable, Agro-horti, Agri-silvi-pastoral models of different combinations of suitable, identified multi-use-tree species and agriculture crops for different agro-forestry Systems and agro-climatic-zones, as well as selection of suitable indigenous multi-purpose tree species for Farm Forestry models, value addition in the timber, fodder production, resource inventory, value addition in the NTFP through mechanism of collection, storage, grading and marketing etc.

Research studies are also needed in the field of protection as well as, production forestry i.e. protection of naturally, as well as, artificially regenerated stands, seeds and nurseries from fungal diseases, insects and rodents` pests, etc. and to take maximum yield through bio-mass production etc.

No research is complete until the research findings are accepted and adopted by the users. For dissemination of research results, the research findings need to be properly documented in journals, newsletters and pamphlets. The developed and tested technologies should be transferred to the users for efficient use in the field.

As on today, there has not been much research extension work. Enough potential lies unexplored. Sky is the limit in this field of forestry research and it's extension. But due to, inter-alia, inadequate infrastructure facilities, insufficient and unwilling field staff and inadequate availability of funds are main constraints in this development process.

(2) Thrust Areas for Forestry Research and Extension

The following are the important areas of the forest conservation and management, which require the research study, for measures to be taken for scientific and sustainable management of the resource in the changing scenario. Planning and orienting the research for increasing forest cover, developing technologies for conservation and protection forestry, increasing bio-mass production and bringing greenery to the denuded Aravallis, hostile and immortal sand dunes and arid and semi-arid lands with salt affected problematic soils is a gigantic task. Forestry research should develop technology to generate stable, holistic models based on ecosystem dynamics

The researches should be conducted precisely to the micro-forestry-zones in all over the states of the country and research findings should be communicated to the Extension Units, proposed to be created at regional level, for the very purpose.

- **Conservation and Protection**

The fields which need immediate research studies are JFM consolidation, success of the participatory management, follies, proof-ness, usufruct sharing, Gene-pool conservation studies etc. Protection of *Khejari* and *Babool* trees from insect pest, hollowing of *Rohida* trees due to fungal infection etc. and similarly other growing tree stock need proper surveillance and immediate control measures.

- **Production**

To achieve sustainability level, the fields need immediate research studies and research findings to develop suitable models to improve soil fertility, productivity and reclamation of saline, alkaline and acidic soils and waterlogged areas, improvement in teak quality, biomass production enhancement in Bamboo areas, standardization of techniques, right from seed collection to plantation raising, etc.

(3) Non Timber Forest Products (NTFP) and medicinal plants

NTFP and medicinal plants are valuable economic reserves, both in terms of income and employment potential. Approximately 5.00 million tribal people sustain themselves through the collection, processing, transporting and marketing of the forest produce, supplementing their meager income from harvest. NTFP amounts to nearly 80% of the total forest revenue of the state. Present collection system is unscientific because of lack of proper resource inventory, scientific method of collection, technical know-how in processing, storage, grading, as well as, marketing, which has also hampered the development of NTFP based cottage industries in tribal areas. Besides plenty of Medicinal plants, about 45 types of NTFP are found in the natural forests of the state. Scientific collection, processing and value addition is needed.

(4) Soil and moisture conservation

Emphasis has to be laid on standardization of silvi-climatic zones (Agro-climatic zones) specific *in-situ* soil and moisture conservation structures.

(5) Agro-forestry

An integrated land use pattern includes Farm Forestry, Agri-siviculture, Silvi-pastoral, and Horti-silvi-pastoral systems- the most self-sustainable models to produce tree crops together with agricultural crops for maximum economic and ecological benefits: cultivable wastelands, marginal and sub marginal lands to optimal productivity through agro-forestry systems and selection of Multi-Purpose-Tree-Species (MPTS), which are compatible with arable farming and can be introduced in various Agro-forestry systems.

(6) Utilization

Utilization of Forest Produce needs research studies in harvesting, transportation, conversion, processing and storage and marketing field.

(7) Participatory Sustainable Management

At present, about 4000 VFPMCs are working in the state. JFM consolidation, success of the participatory management, follies, proofness and usufructs sharing mechanism needs further studies.

Forestry Research in the State

Present Status

Forestry research began in Rajasthan in 1960 after creation of the Forest Department in the state. To under take forestry related research, a small wing under state silviculturist was created. Initially, the research was focused on growth studies through a system of sample plots and preservation plots with a view to determine rotation for maximum volume production. The forest research was made broad based after the advent of externally aided projects in the state from the seventh plan period onwards.

The forestry research infrastructure with respect to manpower, building, material and investment has been very low despite the fact that other wings have grown with the expansion of the Forest Department.

At present, one, Deputy Conservator of Forest is working as Silviculturist. One Assistant Conservator of Forests, two scientists on deputation from Agriculture department are working as Research Officer Seed & Soil respectively and five Range Officers assist him with some supporting staff. The Conservator of Forests (Research), Jaipur is supervising the work of silviculture wing.

The state silviculture branch has only the following four-stations/ centers for carrying out forestry research work-studies

- World Forestry Arboretum, Jhalana, Jaipur.
- Grass Farm Nursery, Khatipura, Jaipur.
- Govindpura Research Farm, Jaipur and
- Banki, Research Farm Sisharma, Udaipur.

Since inception of silviculture wing, the various research activities, which have been undertaken so far, are described as below

- Preparation of volume tables of *Anogeissus pendula* on the basis of sample plots.
- Preparation of Weight Tables of ten forest species on the basis of actual data collected from felled forest coupes.
- Preparation of local volume tables of *Eucalyptus camaldulensis, Dalbergia sissoo, Acacia nilotica, Acacia tortilis* and *Prosopis cineraria.*
- Preparation of fodder yields tables of *Prosopis cineraria* and *Ailanthus excelsa.*
- Adaptive research and on farm trails.
- Collaborative research programmes with institutions like Arid Forest Research Institute (AFRI), Jodhpur, Tata Energy Research Institute (TERI), New Delhi and State Agriculture University (SAU), Udaipur.
- Marking of Candidate Plus Trees of eleven important species.
- Provenance and introduction trails of exotic and indigenous tree species.

On going research activities

With the limited available staff and research facilities at present, following research activities are being carried out

1. Provenance and introduction trails of the following important indigenous species are being carried out at research center Govindpura Research Farm, Jaipur:
   - i. Sisham (*Dalbergia sissoo*),
   - ii. Eucalyptus (*Eucalyptus camaldulensis*),
   - iii. Babool (*Acacia nilotica*),
   - iv. Ardu (*Ailanthus excelsa*),
   - v. Khejari (*Prosopis cineraria*) and
   - vi. Rohida (*Tecomella undulata*).
2. The studies of nursery techniques for indigenous species being carried out are:
   - Hingot (*Balanites aegyptiaca*) and Ker (*Capparis decidua*): at Grass Farm Nursery, Khatipura, Jaipur.
   - Sitaphal (*Annona squamosa*) and Malkangani (*Celastrus paniculatus*): at Research Farm, Banki, Shisarma, Udaipur.
3. Biomass production study of Bamboo in Torna Forest Block of Udaipur Forest Division.
4. Use of Bio-fertilizers: The studies are being carried out about the impact of use of bio-fertilizers on the growth of *Prosopis cineraria* seedlings by using potting mixture of *Mycorrhyza, Azatobactor* and PSB bio-fertilizers in different combinations for raising seedlings in the poly pots in the nurseries.
5. Study of vegetative propagation of Guggal (*Commiphora wightii*) in regulated temperature and humid conditions in mist chamber using hi-tech nursery techniques. Study has been concluded, as humid climate is not suitable for it's vegetative propagation.
6. Collection of seeds of the following indigenous species from identified Seed Production Areas (SPAs):
   - Kumatha (*Acacia senegal*): from Pushkar, Ajmer district.

- Khejari (*Prosopis cineraria*): from Manda Bhainslana, Jaipur district.
- Sisham (*Dalbergia sissoo*): from Norangdesar Distributary, Hanumangarh district and
- Babool (*Acacia nilotica*): from Gharsana and Anoopgarh area, Sriganganagar district.
- The study carried out about yield of clean seeds of ratanjot (*Jatropha curcas*), reveals that average 2.5 Kg clean seeds can be obtained, every year, from fifth year onwards from an average sized ratanjot tree.
- Testing of seed germination and reporting the results, as and when seed samples are received from the Forest Divisions.

7. The study carried out on the tissue-cultured plantlets, procured from Delhi and transplanted at World Forestry Arboretum, Jaipur during 1998, regarding vegetative propagation of *Anogeissus pendula* through tissue culture revealed that it is not successful.
8. Testing of soil and water samples and reporting the results, free of charge, as and when the samples received from the Forest Divisions, as well as, from individual Farmers and other agencies.

Future Strategy

Research to improve forest management haş been a mainstay of forestry science. In view of the recent transition of society's environmental expectations, forest science has to address the functional linkages between people and forests. In order to ensure sustainable development, forest management should develop technology to generate stable, holistic models based on ecosystem dynamics. These will focus on key interactive biophysical and socio-economic elements of forest performance leading to improved cost efficient management options.

In order to implement forestry research, the Forest Department will adopt following strategy:

1. Strengthening of the existing research set up in the state and infra structural augmentation.
2. Prioritization of research problems and identification of research agenda for short-term and long-term perspective and allocating responsibilities.
3. Dissemination of research results through proper documentation vide periodic news letters, annual reports and database creation.
4. Technology extension.

(i) Strengthening of Infrastructure

Presently, there is only one Silviculture Division in the State Forest Department to address the research problems of the entire state. Looking to the peculiarities of edapho-climatic zones in the state, post of silviculturist, are proposed to be created.

The work of the silviculturists should be further strengthened by availing the services of Biometrician, Ecologist, Statistician, Economist, Forest Geneticist, Seed Analyst, Agrostologist, System Analyst, Computer programmer, Sociologist/Anthropologist and Documentation Officer.

(ii) Identification of Research Problems

Through periodic dialogues with practicing foresters, private forestry entrepreneurs, community woodlot managers, VFPMCs, NGOs and forest based industries, research problems shall be identified in a participatory process. However, the specific areas, which will be addressed, are as under:

1. Standardizing micro-planning methodology and resource inventory for meeting JFM needs.

2. Development and transfer of needs based field oriented cost effective technologies.
3. Development of planting technology, use of seeds, clones and tissue culture techniques for promoting improved varieties of indigenous species.
4. Standardizing low cost silvicultural options for multiple product forestry.
5. Rationalising silviculturally sound lopping practices.
6. Undertaking studies to assess socio-economic and environmental impact of the forestry development programmes.
7. Studies of resource assessment particularly of the availability of NTFP including medicinal plants of economic importance.
8. Undertaking marketing studies on demand and supply situations of various forest produce.
9. Identification and adoptation of indigenous knowledge.
10. Undertaking studies on population dynamics to assess the optimum carrying capacity of wildlife habitats, as well as forests, pasture lands and other ecosystems.
11. Undertaking studies on endemism biodiversity outside PAN.
12. Undertaking studies to assess impact of management intervention in PAN and JFM areas.
13. Undertaking studies relating to changes in the wildlife habitats.
14. Research on *ex-situ* and *in-situ* conservation linkages, translocation of captive bred animals.

Research problems identified shall be prioritized and placed before the Forestry Research Advisory Committee (FRAC) for approval of research agenda for a specified time frame. The FRAC members shall include representatives from AFRI, CAZRI, SAU, etc. to avoid duplication of research efforts. For the research to be carried out by the national level institutes, NGOs and SAUs in a collaborative manner, required MOU will be drafted by clearly specifying the responsibilities of the partners. Agroforestry related research shall be entrusted to the Bikaner Agriculture University focusing on the arid zone.

(iii) Documentation of Research Output and Transfer of Technology

No research is complete until the research findings are accepted and adopted by the users. For dissemination of research results, the research findings need to be properly documented in journals, newsletters and pamphlets. Developed technologies need to be transferred to the field for efficient use by the users. At present, even though there is no established extension practice, the forest department is using it's own extension services along-with the services of other line departments. There is an urgent need for preparation of database. For wider communication, the department has launched it's own web site.

(iv) The Research priorities of the State

Besides the broader framework of research as stated above, emphasis will be laid on the following:

1. Seed collection, Storage, certification and distribution after proper testing of germination percentage, plant percentage and treatment methods. These activities will be extended to all seeds used in the state for natural forests, as well as, man-made plantations.
2. Genetic improvement of germplasm through the identification of phenotypically superior trees and stands for collection of seeds for raising superior planting stocks. Since both provenance and progeny trials are time consuming processes, wherever clonal propagation is possible, clonal seed orchards will be

established quickly without waiting for establishment of seedling seed orchards. In case, vegetative/ clonal propagation methods are not standardized, tissue culture plantlets will be used.

3. Species to be given preference for improvement in a reasonable time frame for Rajasthan are as follows:

Short term

1. *Acacia nilotica*
2. *Acacia tortilis*
3. *Prosopis cineraria*
4. *Zizyphus spp.*
5. *Prosopis juliflora*
6. *Dalbergia sissoo*
7. *Ailanthus excelsa*
8. *Eucalyptus camaldulensis*
9. *Azadirachta indica*
10. *Tecomella undulata*

Long term

1. *Anogeissus pendula*
2. *Diospyros melanoxylon*
3. *Boswelia serrata*
4. *Albizzia odoratissima*
5. *Salvadora oleoides*
6. *Acacia senegal*
7. *Acacia leucophloea*

This work will be taken up by Silvicultural wing of the SFD in collaboration with AFRI, CAZRI, SAU, TERI, Rajasthan University, etc.

4. Production studies in plantation models by taking up studies on various sites of the state to assess actual production level with wide varieties of

management interventions. It will be undertaken in existing areas of natural forests/man-made plantations.

5. Carrying out work-studies for different forestry operations covering all forestry activities starting from forest nurseries, plantations, after care and harvesting of the product.
6. Biometrics studies for estimating the harvestable products from tree species, correlating them with easily measurable growth parameters of tree through mathematical modeling by preparing local volume table, fodder yield table, etc., for various species and eventually fit them in a wider area general table. Total biomass assessment will also be done.
7. Evaluation and standardization of *in-situ* soil and moisture conservation devices such as Contour dykes, Contour trenches (both staggered and continuous), Box trenches, V-ditches, Ring trenches, Gradonies, etc., to facilitate natural growth of plants and for ensuring better survival of planted seedlings.
8. Operational research to standardize optimum irrigation, water conservation in arid areas, fertilizer application etc. in plantations will be carried out. Similarly use of water imbibing resins/ hormones, anti termite mulching, weeding and hoeing, etc., will be studied for their optimal application looking to peculiar arid state of this region.
9. For agro-forestry, suitability of species, spacing, intercropping, species mix, time of planting, application of fertilizer/pesticides, farm budget analysis, synergistic behaviour of trees and agriculture crops will be studied and on-farm trials adopted.
10. For promoting cost-effective technologies, studies will be carried out for various types of fencing, pit sizes, nursery trials with different sizes of poly pots;

poly pot vs. bare roots plantings; grazing vs. grass cutting; grazing vs. stall feeding; root training; raising of tall plants, use of fuel saving devices; Ethology of CPR management; benefit/usufruct sharing methods; value addition and marketing; Tree planting as a mode of drought proofing and economic security; etc.

11. Some of the newly emerging areas like valuation of forests, role of trees in pollution abatement, waste-water utilization through tree planting and carbon sustainable management. The state research wing will not be required to undertake research in these fundamental areas. However, research findings of the national and international forestry research institutes will be made use of as and when required for planning and management purposes.

Forestry Extension

(i) Need assessment

Beginning of Forestry Extension was made with the advent of externally aided National Social Forestry Project from the year 1985. Though the idea was well conceptualized and to a limited extent agriculture extension network was also used to promote forestry extension but it could not be continued systematically due to a variety of reasons. It could only get momentum during the implementation of Japan assisted Aravalli Afforestation Project and Forestry Development Project in 90s. The extension activities remained limited to the project areas only.

(ii) Present status

At present, one unit, headed by Range Forest Officer is working at state forestry headquarters and another headed by a DCF is working at Rajasthan Forestry & Bio-diversity Project's headquarters, with bare minimum supporting staff;

i.e., one RFO, two Foresters and two Forest Guards with huge jurisdiction, comprised of eighteen districts of the state. Therefore, it is very essential to have a proper forestry research extension network so as to promote forestry development in the state.

Future Strategy

(i) Infrastructure

It has been proposed to essentially set up independent extension units, with necessary supporting staff and equipments such as EDP, Audio-visual aids and vehicles at the regional levels, *viz*, at Jaipur, Jodhpur, Udaipur, Kota and Bikaner under the regional CCFs. At the state level, a 'Network' will be established between all the regional extension units so as to allow standardised database generation, documentation and methodology.

(ii) Awareness raising

Awareness raising about the use of research findings is extremely pertinent in relation to forestry development in a third world country like India. Extension units set up as above will apply the following methodology for raising awareness:

1. Proper documentation

Extension units will prepare, develop and disseminate publications, pamphlets, posters, stickers and any other relevant publicity material. Setting up of libraries and related documentation facilities etc.

2. Organizing film shows, exhibitions, chaupals, puppet shows etc.

By organizing film, video shows, puppet shows, chaupals, exhibitions, etc. both for urban and rural populace. Such activities will be specific target group oriented.

Maximum use of folk art forms will be made. Extension services will be extended to schools and educational institutions.

3. Experience sharing workshops and exchange visits

In order to ensure bi-directional feedback, the extension units will be responsible for organizing exchange programmes and workshops for all categories of forestry personnel, extension workers, VFPMC members, Eco-Development Committee members, Local Self Government representative etc. In the normal course, exchange visits will be followed by a workshop where sharing of field experience and cross-fertilization of ideas will take place.

4. Use of mass media

The importance and capabilities of mass media as a tool for extension is acknowledged. The extension units of the SFD will liaise with and supply to the mass media information and software so as to allow maximum publicity. These extension units will handle TV programmes, radio talks, press release, etc.

5. Sensitization of Politicians, Public servants, Media personnel etc

The need to sensitize politicians, public servants and media persons to the needs, requirements and performance of the SFD is of paramount importance. They need to know, appreciate and, when necessary, critically analyse the forestry activities. For this purpose, the extension units will be cultivating this very vital target audience by preparing reading material, literature consulted, training and will also organize field visits from time to time so that interaction among field functionaries, locals, other departmental staff would take place and first hand knowledge of the methodology adopted would also be available.

(iii) Promoting JFM including proper functioning of VFPMCs and Eco-development Committees

Joint forest management in wildlife and multiple use areas calls for substantial extension activity. The extension units in collaboration with local divisions will ensure proper functioning of CPR management committees by providing training and need/situation-based software. Identification of local resource persons amongst the committee members will be done who will serve as extension workers after being trained. Provisions for honoraria will be kept in the extension unit's budget. Both eco-development committees and VFPMCs will come under these units. Proper training, monitoring and feedback in consultation with the locals will ensure active participation. Use of local knowledge and socio-religious traditions of the resource conservation practices would not only help in getting active participation and co-operation of the local people but also help in confidence building.

(iv) Use of Agriculture Extension Network specially for promoting Agro-forestry/Farm Forestry

The Extension wing of Agriculture Department is extremely well equipped. Their grass root level network will be used regularly in awareness raising, especially with regards to the benefits of Agro-/Farm forestry. However, these workers will first be put through orientation courses so as to sensitize them about the modalities and needs of Forestry development. The officials of the extension units of SFD will also regularly participate in the training, workshops and field visits organized by the extension wing of the Agriculture Department. The Forest Department extension units will liaise with the agriculture extension wing.

Constraints in Forestry Research and Extension

- Need for well planned field oriented research plan.
- Inadequacy of infrastructural research facilities.

- Inadequacy of funds.
- Inadequacy of dedicated staff.
- Need of decentralized regional level research centers.
- Activation of forestry awareness centres (Van Chetana Kendras).
- General apathy of the villagers (users) and lack of appreciation of role and value of forests and wildlife.
- Lack of documentation and proper record keeping of research findings for extension.
- Lack of network system for forestry extension.

Suggestions and Recommendations

- A changed mindset is desirable.
- Proper emphasis on research and extension in forestry projects.
- Creation of independent forestry extension units, at regional level with adequate infrastructural support.
- Availability of adequate funds.
- Non-existence of Research Extension Wing.
- Deployment of willing & dedicated man power in research wing.
- Incentives for outstanding research workers.
- Creation and regular running of Forestry Awareness Centers (Van Chetana Kendras) in sufficient numbers scattered through out the state. At least one in each district at district headquarters in the campus of Central Nursery.
- Decentralized regional level research centers.
- Establishment of easily approachable Demonstration Plots through out the state, in sufficient numbers.
- Human Resource Development. Capacity Building by organizing trainings programmes, workshops and

field visits of the field staff as well as the user groups.

- Regular updating and upgrading of literature, exhibits and information related to forestry research out comes and developed technologies.
- Micro-level Forest Zonation (Silvi-Climatic-Zones) from forest vegetation point of view on the lines of Agriculture department's Agro-climatic Zones for proper research & development of forest resources.
- Regular collection and documentation of indigenous and Ethno-forestry knowledge.
- Research Findings must be translated into local language for easy dissemination and transfer to the users.
- Target group specific Communication and Extension strategy.
- Regular Periodical Review of the Research and Extension work.
- Networking of Forestry Extension Units with Research Institutes.

Literature consulted

Agarwal, S.R. 1981. Trees, flowers and fruits in Indian folk songs, folk proverbs and folk tales. *In*: S.K. Jain (ed.) *Glimpses of Indian Ethnobotany*. Oxford and I B H Publishing Company, New Delhi, pp. 3-12.

Bahuguna, V.K., Kinsuk Mitra, Doris Capistrano and Sushil Saigal. 2004. Root to Canopy: Regenerating Forests through Community – State Partnerships Multipurpose Tree Species For Agroforestry in Utter Pradesh, Rajasthan & Haryana, 1995 – Regional Centre, NAEB, Agricultural Finance Corporation Ltd., Bombay.

Bhandari, P.N. and Chaurasia, N.K. 1981. Farm Forestry in tribal area of Rajasthan. *In*: N.N. Vyas (ed.) *Social Forestry in Tribal Development*. *Tribe* Spl. No. 13(2-4): 49-60.

Forestry Development Project. 1995-2002. *An Overview*, Forest Department, Rajasthan, India.

Forests Department, I.G.N.P. 2001. Afforestation and Pasture Development along Indira Gandhi Canal – *A Commitment for Greening of Thar* (March, 2001) – Forests Department, I.G.N.P., Bikaner.

Gandhi, Tara. 1991. *Rajasthan Vanspati Abhisuchak,* Parati Vikas Samiti, New Delhi.

Joshi, P. and Awasthi, A. 1992. Tribal devices and techniques against pests and farm enemies of Rajasthan.

Joshi, Prabhakar, 1995. Ethnobotany of the Primitive Tribes in Rajasthan.

Kapoor, R.P. 2005. Key Note Address: Workshop on Grassland Ecology & Gene-pool Conservation, Sept; 27-29, 2005, Jaipur.

Lal, J. B. 1992. *Forest Ecology*. Nataraj Publishers, Dehradun.

Pandey, Deep Narayan. 1992. Beyond Vanishing Woods.

Rajasthan Forestry Action Plan, 1996-2016. Forest Department, Rajasthan.

Shetty, B.V. and Singh, V. 1987-1993. *Flora of Rajasthan: Flora of India, Series 2.* Botanical Survey of India, Calcutta. 3 Vols.

Shiva Vandana, Jafri Afsar H. and Gitanjali Bedi. Ecological Cost of Economic Globalisation: *The Indian Experience.* Research Foundation for Science, Technology and Ecology.

State of Forest Report. 1999. Forest Survey of India (MoEF), Dehradun.

State of Forest Report. 2001. Forest Survey of India (MoEF), Dehradun.

State of Forest Report. 2003. Forest Survey of India (MoEF), Dehradun.

The Indian Forester, Environment Special, June, 1991.

The Indian Forester, July 2000 – Vol. 126.

The Indian Forester, June 2005 – Vol. 131.

The Indian Forester: *Special Issue on Wildlife Management* October, 1986 – Vol. 112.

Verma, S. K. 1991. Rajasthan Ka Vanya Jeevan.

EXTENSION TECHNOLOGIES FOUND USEFUL IN TREE IMPROVEMENT WORK

C.J.S.K. Emmanuel, U.K. Tomar and Tarun Kant

Division of Forest Genetics and Tree Breeding,
Arid Forest Research Institute, Jodhpur

ABSTRACT

Tree improvement activities in India were catalyzed during the year 1962 following the visit of Dr. J.D. Matthew to India who submitted a report emphasizing the need to take up research in tree breeding to fulfill the needs of quality seed. Major emphasis was placed on a long term breeding programme for teak with the object to produce superior varieties or cultivars suitable for growth on the moist, semi-moist and dry sites. Later in the year 1975, a project on Tree Improvement and Seed Procurement was started in collaboration with the Government of Denmark which provided much needed field orientation to the tree improvement activities in India. Then the assistance was provided for this work by UNDP and lastly by the World Bank.

Lot of work was done in collaboration with the State Forest Departments for which training was essential. Initially, short term courses were started for the forest officials by the Forest Genetics Division, Forest Research Institute, Dehradun by Dr. S Kedarnath. A demonstration clonal seed orchard for teak was established at the Forst Research Institute. In the Indo Danish Project on Seed Procurement and Tree Improvemenmt financial assistance, in the form of equipment and vehicles was provided to five State Forest Departments to open Seed Certification cells. Guidelines were prepared for the selection of Seed Stands and their conversion into Seed Production Areas. Guidelines were also prepared for the selection of Candidate Plus Trees and their conversion to Plus Trees.

To carryout the work efficiently, in every state of India a Nodal Officer from the State Forest Department was selected. Meeting of these Nodal Officers was conducted every six month, in which a representative of DANIDA and a representative from the Ministry were members, to monitor the progress of the work and to fix targets for the next six months.

To keep track of the work Syscon Visualisers were distributed to every State Forest Department for recording the data on Seed Production Areas, Plus Trees and Seed Orchards. A copy of these records was also available with the regional centers of Indo-Danish Projects.

To help in conducting the work, the project has delineated the provenances with the help of Conservator of Forests (Research) in each State and also prepared the Forest Seed Certification Scheme. The present paper discusses all these aspects and their success in implementation through extension practices.

Introduction

To raise large plantations of economic value, it is essential to use seeds which are genetically improved for desirable traits. Forest tree breeding has the highest potential for maximizing yield per unit area and time along with bringing improvement in the quality of forest products from plantations. Significant achievements have been reported from many countries, both, in conifers and broadleaf species, in terms of higher productivity, i.e., improved growth rate and yield; wood properties, resistance to insects pests and diseases. However, more work remains to be done on developing improved strains suitable for growing in adverse environments or areas. To develop genetically improved quality of planting material, it is essential to define the objective for which improvement has to be brought. Though the objectives may vary depending on the choice of species as well as the end use, yet some of the objectives may, in general, include; breeding for adaptability to different environments and soil conditions, timber production, wood properties, resistance to diseases, minor forest products, agro- forestry etc.

Tree improvement activities in India were catalyzed following the visit of Dr. J.D. Matthew to India who submitted a report emphasizing the need to take up research in tree breeding to fulfill the needs of quality seed. Major emphasis was placed on a long term breeding programme for teak with the object to produce superior varieties or cultivars suitable for growth on the moist, semi-moist and dry sites (Kedarnath and Matthew 1962).

Later in the year 1975, a project on Tree Improvement and Seed Procurement was started in collaboration with the Government of Denmark which provided much needed field orientation to the tree improvement activities in India (Anon, 1979).

In genetical improvement of tree species major emphasis is placed on selection breeding. The major significance of selection is that, parents are chosen for the next generation. Selections are always made in view of the economical importance of the traits or characters. Following points are generally considered while choosing traits for selection purposes.

- Sufficient genetic variation for the character should be available,
- The character should be heritable and its heritability should have been worked out,
- Trials have been conducted to screen out geographic variation available in that species.

Selection can be either on individual trees or those of good plantations for seed collection. To get improved quality of seeds, two types of programmes are generally followed:

1. Short term programme-which includes seed stands and their conversion to seed production areas for quick supply of quality seed.
2. Long term programme which includes selection of plus trees, establishment of seed orchards, laying out of progeny trials, selection of elite trees, establishment of second generation seed orchards etc., but it takes a long time to give quality seed material.

Approach and Methodology

Seed Production Area

Seed production area is referred to as a phenologically superior stand made up of vigorously growing healthy trees, upgraded by thinning to remove poorer phenotypes, treated and managed to cause abundant seed production. Since the culling of inferior trees is recommended, it is also called as negative selection, because we select inferior phenotypes for removal from the seed stand. The genetic gain obtained through selections is 5 to 10 percent.

Establishment of Seed Orchards

Selection of plus trees

Plus trees are individual trees of out standing merit, initially selected on the basis of superior phenotypical characters like height, diameter, clear bole, etc. Selection of plus trees involves:

- Preliminary reporting of the out standing tree called candidate tree,
- Final phenotypical appraisal and approval as plus tree.

Preliminary reporting

The preliminary reporting of the candidate trees serves the purpose of screening as large areas as possible. This initial search for candidate tree can be done even by the junior staff, which is more familiar with the field crop within their jurisdiction. The criteria for screening the trees are as follows

The final phenotypic appraisal is done by the Forest Geneticist / Silviculturist and other members of tree improvement group. So, that the candidate's superiority or the phenotypic expression is critically analyzed and discussed with out any bias going in favour or against the selection. The height, clear bole and diameter are the actual measurement taken but for the other characters scoring is done with the help of ocular estimates.

Evaluation of these trees is carried out depending on the location of candidate tree whether in plantation or in natural stand. There are basically two methods applied for the selection:

- Comparison tree selection.
- Base line selection.

The comparison tree selection or check tree selection is commonly used in even aged stands for a species (Cech, 1959; Pitcher and Doru, 1967). The first step is the selection of the candidate tree. Once the candidate trees are selected they are screened for traits of interest in relation to few surrounding trees called comparison tree or check tree. If the candidate tree exceeds the comparison tree, it is selected as Plus Tree. The objective behind comparison tree is to adjust or correct the phenotypic value of the candidate tree for environmental effects common to that particular stand. It is presumed that environmental differences between stands or areas that vary in soil, climate or stand history would make it difficult to evaluate breeding value on the basis of phenotype. The environmental check through the use of comparison trees is believed to result in an improvement in the accuracy of recognizing individuals with good genotypes and not merely good phenotypes.

While selecting plus tree in an uneven aged stand with mixed species growing, the method adopted is the base line selection (Beineke, 1969). In this method the individuals are located and their value for traits of interest is compared to the average of the region in which the selections are made. The average is a "base line" giving the system its more common name" base line selection". If the candidate tree exceeds the base line by a considerable amount, it is selected and incorporated in the breeding population. The base line may take the form of a regression equation relating height or diameter to age but it could even be a multiple regression equation that takes into account physical factors of the site. The candidate tree in this case is not compared to surrounding trees.

Seed orchard

A seed orchard is a plantation of selected clones or progenies which is isolated or managed to avoid or reduce pollination from outside sources and managed to produce frequent, abundant and easily harvested crops of seed (OECD, 1974).

The concept of seed orchard started with the use of quality seeds as suggested by F.A.L Burgsdorf of Germany in 1787, using specific plantations raised through vegetative propagation in forest tree breeding. Establishment of seed orchards converted from seedlings of known seed origin was suggested in 1918 by N. Sylven. In 1931, first forest tree seed orchard was established in Scotland by J. Scrymgeour Wedderaburn. This method of tree improvement was taken up by most of the western countries, and later on by all over the world for the production of quality seeds.

Though it is long term quality seed production programme but the expected genetic gain from such orchards are from 30 to 40 percent. The advantage of this breeding system can be obtained if the future plantations are raised from the seeds produced by such orchards. Seed orchards are of two types:

1. Seedling seed orchards.
2. Clonal seed orchards.

Seedling seed orchards

Seedling seed orchards are raised from the seeds collected from superior phenotypes. The seedlings planted from this material are roughed to remove inferior trees, leaving the best trees of the best families for seed production.

Clonal seed orchards

A seed orchard composed of vegetatively propagated trees established primarily from the clonal material of superior phenotype for the production of seed of proven genetic quality.

The stated goals of seed orchards are to produce genetically improved seed of forest trees, to promote frequent seed crops, and to facilitate harvesting. On an International scale, specific rules on minimum standards for tree seed orchards have recently been published, as guidelines for countries particularly in the OECD Scheme for the Control of Forest Reproductive Material Moving in the International Trade. Although these rules specify that self pollination and outside pollen contamination should be minimized, they do not give any quantitative guidelines or establish any specific requirements. The responsibility of defining specific criteria for seed orchard belonging to the various categories recognized in the scheme is left to the national authorities responsible for its implementation.

If the concept of genetically improved seed is to be respected, some guarantee for the composition and properties of seed lot must be provided. If possible legislation requiring official registration of seed orchard and other forest reproductive material should be evoked to check inferior planting material.

Seedling seed orchards

Though in general the quality seed production in forest trees is limited to clonal seed orchards, but the suggestion given by N. Sylven in 1918 regarding conversion of seedlings of known seed origin into seedling seed orchard has laid down the foundation of seedling seed orchard. When compared to clonal seed orchard, these orchards require a bigger large area, more number of trees for block planting as families and it takes longer duration for seed production.

The seed produced by these seed orchards fall in two categories of seed certification scheme. The seed collected from first generation seed orchard comes under the category "Reproductive material collected from untested seed orchards", and seeds collected from 1.5 and 2nd generation seed orchard comes under the category of "Tested Reproductive Material ".

Inference

The most important thing which was found useful is the interaction of the forest officials in planning and implementation of the tree improvement work. During the DANIDA project, the monitoring and planning were given prime importance. To facilitate this DANIDA project was conducting six monthly monitoring and planning meetings. This meeting was attended by all the State Forest Department Liasion Officers. The Liasion Officers were Conservator of Forests (Research), the targets were fixed with their consent and later on monitored with their assistance.

Similarly in the World Bank project also the Silviculturist of the State Forest Department was mad the Nodal Agency for carrying out the work and all field activities were done through his staff. Whenever and wherever it was needed the staff was given the training for all activities conducted for tree improvement work.

Literature consulted

Barber, J.D. and Dorman, K.W. 1964. Clonal or seedling seed orchard? *Silvae Genetica* 13 (1-2): 11-17.

Beineke, W.F. and Lowe, W.J. 1969. A selection system for superior black walnut trees and other hardwoods. In: Proceedings of the Tenth Southern Conference on Forest Tree Improvement, Houston, Texas, pp: 27-33.

Cameron, J.N. and Kube, D. 1980. Management of seedling seed orchard from plus trees, Biometrics Team, Agricultural Research Council of Central America, Southern Rhodesia.

Cech, F.C. 1959. Superior tree selection-a comparision of gradeing systems, in Proceedings of the fifth southern conference on forest tree improvement, Aleigh, North Carolina, pp: 49-52.

Emmannuel, C.J.S.K. Kapoor, M.L. and Sharma, V.K. 1990. Achievements of tree improvement and their significance in gene conservation, *J .Eco. Bot. & Phy*. 1: 48-54.

Kedarnath, S. and Matthew, J.D. 1962. Improvement of teak by selection and breeding, *Ind. Forester,* 88: 277-284.

Kedarnath, S. Rawat, M.S. and Chauhan, V.S. 1970. Early growth performance of twenty clones of teak (*Tectona grandis* L.f.) in a seed orchard, Proc. Seminar on the Genetic Improvement of Forest Tree Seed in India, Forest Research Institute, Dehra Dun, pp: 86-89.

Klein, J.T. 1974. "A Jack Pine seedling seed orchard plantation of unusual design". 21st North-east *For. Tree Improv. Conf. Fredriction*, New Brunswick , pp 55- 65.

Matthew, J.D. 1961. A programme of forest genetics and forest tree breeding, Research Report to Govt. of India, reproduced for FAO Publication No. 1349. OECD (1974) OECD Scheme for the control of Forest Reproductive Material Moving in International Trade. Organization for Economic Cooperation and Development, 2 rue Andre Pascal, 75016, Paris, France.

Pitcher, J.A. and Dorn, D.E. 1967. A new form for reporting hardwood superior tree candidates, In Proc. of the fifth Central States Forest Tree Improvement Conference, Wooster, Ohio, pp: 7-12.

Riemenschmeider, D.E. 1977. "The Genetic and Economic Effects of Prelininary Cuttings in the seedling orchard". Proc. 13 th

Lake *States For. Tre Imp. Conf. St. Paul Minn., U.S. Forestry Service Technical Report* N. C. 50 st 81- 91.

Thapliyal, R.C., Vakshasya, R.K., Uniyal, D.P. and Rawat, M.S. 1988. Tree improvement activities in Himachal Pradesh. I-Selection of Chirpine (*Pinus roxburghii*) stands and some suggestions for followup action. II-Collection and testing of Kail (*Pinus wallichiana*) seeds. Paper presented in "Interaction with H.P. Forest official" Shimla.

Toda, R. 1964. Vegetative and seedling seed orchard, Special issue of *Silva Genetica* 13 (1).

Venkatesh, C.S. and Arya, R.S. 1967. Observations on the flowering and fruiting behaviour of semul grafts. *Ind. Forester* 93: 586-587.

Wright, J.W. 1962. Improvement rate through clonal and seedling seed orchards, *Proc. fifth world For. Cong. Seattles.*

INFORMATION TECHNOLOGY FOR FORESTRY RESEARCH EXTENSION: POTENTIALS, IMPEDIMENTS AND DIRECTION

A.K. Sinha, R.L. Srivastava and U.K. Tomar

Arid Forest Research Institute, Jodhpur-342 005

ABSTRACT

The extension of research in forestry sector has become an important issue today when we have reached to a stage where lot of research information are to be disseminated to the stakeholders in order to benefit them from the forestry research. Steps have already been initiated by different agencies to extend the research outputs to the stakeholders by means of print media, celluloid media like television, CD's etc. and by creating different task groups to disseminate the research to the stakeholders.

In the recent past, Information technology has emerged as the most powerful tool to exchange the information between the people. *The Information technology deals with how to gather, maintain and retrieve the information in an organized way so as to provide the information on the desktop of the users by using the communications and internet technologies.* The information technology is more powerful than any of the other medium of communications like print or celluloid media in several ways.

- There is no time binding as in television media for delivering the information. The information can be obtained any time, the user wants.

- Vast information about any subject can be stored in a centralized or distributed way as required but can be accessed from a single point.
- Information search is very systematic, specific and fast.
- Information retrieval is at the desktop of the user.
- Information flow is bi-directional between experts and users.
- Paperless working helps in conservation of forests.

Lot of work is already going on internationally and nationally to extend the forestry research information by the use of Information Technology to the stakeholders by the way of construction of websites related to the subject. But some more organized effort is needed to make this technology the most viable, comfortable and cheap alternative in order to reach out to the rural and forest plantation areas so that people may benefit from it.

Information Technology is still considered to be the luxury of the urban areas and not the necessity of the rural areas. With the awareness of computers among the youths in villages and mobile and internet technologies already knocking the doors of the rurals, it is very necessary to be ready with the *Information Technology Tools* to be provided to the users and benefit them. The paper highlights the potential of the Information Technology for the forestry extension; the direction of work needed to be done in this area and suggests the solutions for removing the impediments in the national scenario.

Introduction

Forestry research needs to be extended to the stakeholders for its application in the field. Various methods are being adopted in order to disseminate the research results. Some of these methods are:

1. Formation of Task Groups which can physically contact the stakeholders and disseminate the research results in person.
2. Organizing workshops and Kisan-Mela's etc. for the stakeholders.

3. Creating models of the research results for display.
4. Using Print media for publishing research results in the journals, newspapers, brochures, handouts etc.
5. Using celluloid media for display of films from CD's, broadcast of awareness program on Television etc.

Advantages of the Existing Methods

- The advantage of the first three methods is that these methods are highly interactive as the research can be explained in person and the queries of the user can be solved on the spot. If executed properly these methods produce very good results.
- The advantage with the latter two methods (ie. Print and Celluloid media) is that maximum number of persons can be accessed due to the presence of these media's in cities as well as villages now.

Disdavanatges of the Existing Methods

- The disadvantage of the first three methods is that the movement of the task groups takes exclusive time of the manpower. It is very difficult to find the time of experts so that they can move with the task groups in order to explain the research.
- It requires a heavy requirement of funds, vehicles and other resources which makes this method highly expensive. And finally a finite number of persons can be reached using this method.
- The cost involved in the method is proportional to the numbers of persons accessed.
- The disadvantage of the print and the celluloid media is that there is only Uni-Directional flow of information from experts to users. Only experts can disseminate their knowledge to the users but users cannot pose their queries to the experts.

- There is also a binding of time in case of the television programs. The user may or may not find time to watch the program on the specified time.
- The updation of the information in the print media is very difficult and slow. Only limited information can be disseminated using these media's.
- The experts of different fields are scattered and user cannot contact them through these methods on his requirement.

Information Technology – A Powerful Tool

Information Technology (IT) now provides a much better alternative as a powerful media for dissemination of any kind of information. The advantages of using the Information Technology are manifold:

- There is no time binding as in television media for delivering the information. The information can be put by the expert or retrieved by the user from his desktop to the database server through the web-interface any time.
- Vast information about any subject can be stored in a centralized or distributed way but can be accessed from a single point. The information can be stored in multiple database servers and the databases can be accessed by the user through a website from his desktop.
- Information search is very systematic, specific and fast. The user can pose his queries in a very specific way and the database will respond to his queries if the answer of the queries is available in the database. Otherwise the database will store the query itself and will pose these queries to the experts. The expert's reply to the queries will be stored to the database and will be communicated to the asking user through mail.

- Information retrieval is at the desktop of the user having internet connectivity.
- Information flow is bi-directional between experts and users. The users and experts can communicate to each other through the database or can directly form the communication through email exchange or e-chatting.
- Updation and dissemination of the Research Information is very fast as compared to print and celluloid media. The information can be updated through the desktop of the expert by changing the information in the database through the website.
- Exclusive time of the experts is not needed. Experts can feed the response to the queries and give reply to the users queries from their Desktop itself through the web interface. This method becomes very convenient for both the parties and very inexpensive as compared to the traditional methods in the long run.

With all the above advantages over the traditional methods, the Information Technology deserves to be the future method for extending the research results to the users.

Material and Methods

The implementation of the application of the Information Technology for the forestry extension requires some basic hardware, Internet connectivity, Databases, Development platform software and Internet application software as mentioned below:

1. **Hardware Database Server:** At least one hardware server is required for hosting the database server on which all the information regarding the research extension, expert information, user information, user's queries etc will be stored. This hardware server should be a high end server of good

specifications enough for hosting databases like Oracle 9i, SQL Server, MYSQL or any other desired database. Hardware RAID-5 (Redundant Array of Independent Disks) implementation in the server is necessary for the fault tolerance. RAID is technique which employs two or more drives in combination for fault tolerance and performance. Level 5 RAID implementation is the most common implementation for the fault tolerance in the servers. The specifications of the server may be decided at the time of implementation of the project.

2. **Hardware Web Server:** At least one hardware server is required for hosting the website through which the users and experts will interact to the database and to each other. RAID-5 implementation is desirable in the case of web server also. The website may be hosted using Apache webserver on Linux Platform, IIS on Windows Platform etc.
3. **Internet connectivity:** These servers must be provided a continuous Internet connectivity of a very good bandwidth. A leased line connection of 2 mbps bandwidth will suffice to the requirement initially and the bandwidth may be increased depending on the usage of the website by the users. The traffic generated on the website is proportional to the no. of users accessing the website at the same time.
4. **Development Platform:** A front end web application software need to be developed which will be hosted on the webserver and which will serve as an interface (website) between the users and the experts. This web application software should suffice to the requirement of the users in the sense that it should provide the users all the possible options for posing their queries to the database or the experts. This web application software could be developed using Microsoft ASP.NET platform, JAVA platform or PHP platform in combination with the DHTML

pages. This Web application software should provide online connectivity to the database.

The Working Model of the IT application for Extension of Research Results

The Database Server is the backbone of the system which contains all the information to be disseminated to the user. This database can be updated through a front end tool provided to the database updation team. The database updation team fills and updates the information in the database as provided to him by the experts. It operates at the same physical (network) location where the database server resides and accesses the database directly and not through the internet. The database is looked after by a database administrator who is responsible for the overall working of the database server.

The web server hosts the web application software which is used as an interface between the database server and the user. It is better if web server and the database server reside at the same physical location (or LAN). This web application can be accessed through a URL like http://www.extensioninfo.com. The user can access this site by typing this URL in the Internet Explorer on his desktop. This application software will contain a user login page and an expert login page. The new user and experts will have to sign up the website before using the website. As soon as the user or expert signs up, their information is required to be filled up and will be stored in the database. The web application software will contain several forms for entering the queries of the user. The user can pose their queries using these forms. The query will be posted to the database server immediately by the application software which in turn will return the answer to the query if it is available in the database. The answer will then be posted back to the website on a form/report from where the user can see the results. In case the answer of the query is not available on the database, then the query itself will be stored on the database to be posed to the experts for the answers. When, an expert logins the website, immediately the pending queries will be presented to him for answers. All the answers of the queries given by an expert will be stored in the database and will be provided to the user on his next login.

The web application software will have a bulletin board page which will highlight any new research and developments made recently which can be extended to the users. This bulletin board will keep the users up-to-date in the forestry research.

In addition to user and experts communicating to each other through the database, the website will provide direct chatting facility between users and experts. The time of the experts will be decided and put on the website. The user can chat to the expert through this website and directly find the answer to his queries which he could not find through the database.

The System diagram for the implementation is given below:

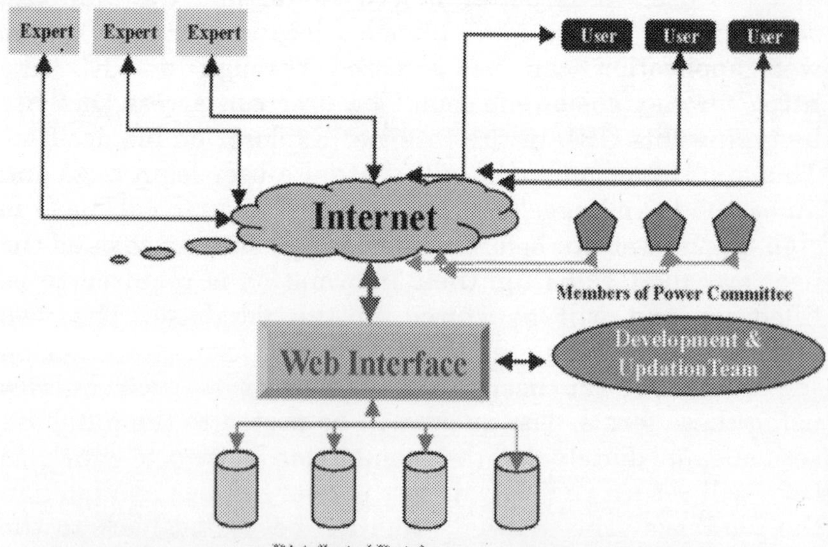

Systematic Diagram for the Implementation of Web Databases

Results and discussions

With so many advantages over the traditional methods of extension, the IT application can prove to be the most powerful tool for extending the knowledge. Some tasks need to be initiated on a high priority basis in order to make the stakeholders ready for the next generation technologies for extension.

- Work should be started in order to enter the existing research information into computerized databases in a standardized format depending on the subject. This format may be decided by the panel of subject experts, IT personnel developing the application software and the users.
- Research should be started to how construct databases of individual areas so that they can also be used to produce combined results.
- The structure of the database should be decided for each area in coalition with other areas and a detailed document should be produced.
- Depending on this document, the construction of web based databases and different front end software tools should be started which can help users to solve their queries using the databases.
- The users should be made aware of these web databases by conducting the workshops right from the development stage of the web database and by advertising it.
- High power committee must be constituted for monitoring the progress of the web-database and for motivating the experts.
- Self help user groups may be formed on internet in order to help each other. That is the web site should also provide a platform where the users may interact with each other and share their experiences.

- The web database should also be made available on the CD with an easy front-end tool for searching the database for those who cannot access the internet.
- The new updates in the database should be made available as patches in downloadable form.
- Small computer center with the Internet facility should be established at the agencies already existing in the villages/forest areas for accessing the database.
- The youth of the villages should be trained and motivated to use these databases.

Impediments to the implementation and solutions

It is a general perception that the computer technology is very expensive and cannot reach to grass root level due to its exorbitant cost. The computer hardware has become cheaper exponentially and it has become a home commodity now a days. If calculated in a long run, the use of IT for extension will prove to be a cheaper option than the traditional methods due to its less recurring cost.

An apprehension in the implementation of this system is that experts may not respond to the queries of the user. If this happens, the whole idea may become defunct. A high pressure committee should be established within the system to motivate the experts and make sure that every query of the user is answered. The experts may also be rewarded for their responses to the users query.

Another impediment may be the non availability of Internet connectivity at remote places. But with the mobile communication technology growing exponentially, very soon internet facilities will reach to the villages/forest areas. Till then CD's of the databases with a front end tool may be developed which can be used on a standalone PC.

As on date, the actual user is not friendly with these technologies due to poor literacy rate. But the literate youth of the village may be trained in using these technologies and self help groups may be formed amongst the users of the same village who can help each other in solving their queries through this database. In order to make this tool effective, front end application software or the website may be developed in the local language of users for higher acceptability and usage. As soon as the people start benefiting from this site, they will accept it and use this to its maximum extent.

Information Technology has already shown its power in the sector of online reservations, Banking, Mobile communications and expert-exchanges of IT related problems. With all these ideas translated to reality, the power of the Information Technology can be harnessed to extend the research knowledge to those who can use it on field.

Literature consulted

SQL Server 2000. Administration /Mark A. Linsenbardt, M. Shane Stigler, Shane Stigler, Mark Linsebardt /October 2000/ McGraw-Hill Professional Publishing.

MCSE SQL Server 2000. Database Design Training Guide/Thomas Moore/March 2002/ Que.

Professional Java Server Programming/ Danny Ayers, Hans Bergsten, Jason Diamond /July 1999/ Wrox Press Inc.

Peter Norton's Guide to Java Programming by Peter Nortan & William Stanek.

Microsoft's Training Material for C # Programming, ADO.NET Programming and ASP.NET Programming.

Professional ASP.NET 2.0 by Bill Erjen, Scott Hansel lman, Farhan Muhammad, Srinivasa SivaKumar, Devin Rader/ Wrox Press Inc.

Orcale Database 10g: The Complete Reference by Kevin Laney/ Oracle Press.

Website Literature Consulted

URL www.wdvl.com: a site detailing how to develop a web database.

URL http://asp.net: a site for development of web-databases in ASP.NET.

URL www.csharphelp.com: a site for development in C#.

URL www.codeguru.com: a site for coding in c# and .NET.

URL http://www.expertexchange.com: a site for exchanging the information for IT related problems. This site matches our idea of creating an extension web database. URL http://www.irctc.co.in a site for online Indian railway reservations.

URL http://Plants.usda.gov Plants Database of USA developed by US development agency. The website provides the information about various species of USA.

URL http://www.plants-india.org: A website developed by the ATREE (Ashoka Trust Research in Ecology and Environment, Bangalore) in association with University of Agricultural Sciences, Bangalore to systematically compile the information on flowering of plants.

DEVELOPMENT OF SCIENTIFIC RATIONALE FOR DISTRIBUTION AND ABUNDANCE OF MEDICINAL PLANTS OF DESERT RAJASTHAN

Manju Singhi, Vinod Joshi and P.K. Dam

Desert Medicine Research Centre, Indian Council of Medical research, New Pali Road, Jodhpur-342 005.
E-mail manju@dmrcjodhpur.org

ABSTRACT

Medicinal plants constitute most important segments of local flora of an area in a desert ecosystem of Rajasthan, India. Availability of these plants carries additional significance for the reason that accessibility of far flung villages is poor and socio-economic level of inhabitants is under developed. An inventory of available medicinal plants in desert, a knowledge of ecological and edaphic factors associated with their occurrence and a programme productive of possible occurrence and abundance of different species of medicinal plants appears to be essential local knowledge which needs to be generated.

In the present paper, an attempt has been made to develop a rationale as to why a particular medicinal plant occur in a particular settings and the same is absent from other settings. Out of 18 plant species studied from selected areas of Jodhpur district, maximum number (12 species) were documented from hilly region, followed by irrigated area (7 species) and least in plain area (6 species). The soil factor of an area emerged to be most important associated factor. The present paper emphasizes on the need to undertake more comprehensive studies on these aspects and a detailed inventory of different plant species of medicinal importance across all ecotypes of desert Rajasthan may be prepared.

Introduction

Medicinal plants form that wealth of an ecosystem which normally caters to the curative needs of the regional ailments of that area. Due to lack of focused and reproducible research to explain the basis of distribution and abundance of flora in a particular region, enough knowledge on the medicinal floral profile of an area is not documented. In the era of availability of advance scientific technology, soil consistency, water chemistry, and stressed physiological conditions encompassing desert flora, can be qualified as well quantified to explain why certain medicinal plants species concentrated and distributed in one area and why are these scarce in other? To study the association of different species of medicinal plants with different ecotype, a pilot study has been undertaken in, hilly, sandy plain and irrigated areas of Jodhpur district of Rajasthan State. We have studied distribution and abundance of eighteen species of medicinal plants namely *Calotropis procera*, *Salvadora persica*, *Capparis decidua*, *Prosopis cineraria*, *Euphorbia caducifolia*, *Acacia senegal* etc. as mentioned in three ecotypes.

In the available literature knowledge about plants of medicinal importance as occurring in Rajasthan is chiefly based on the perceptions of the inhabitant in different regions of the state. About 306 plants of medicinal importance have been reported to be occurring in Rajasthan. Out of these about 100 plants have been documented by desert regions by number of the workers from time to time. Flora of the Indian Desert has been described by Bhandari (1990), Satyavati and Gupta (1987) has given a volume on plants of medicinal importance as occurring in different part of India. Blatter and Hallberg, (1981-21) and Chopra *et al.* (1956)

have compiled chemical and pharmacological aspects of Indian medicinal plants.

In present paper, to test the hypothesis that each species of medicinal plants in desert is dependant on eco-socio factors, work has been accomplished in three different ecosystems *viz*; hilly terrain, plain and irrigated areas as a pilot study.

Presence or absence of a particular medicinal plant has been studied with reference to type of an area and factors governing occurrence have been studied in a demarcated way. The most determining factor of all the parameter studied has been detected and is attempted to be projected as a possible predictor of occurrence of medicinal plants of an area.

Material and Methods

Field trips were carried out in three different geographical regions hilly, plain and irrigated areas of marusthali area of the Jodhpur district as a pilot work. These areas have been screened for the collection of plant species of medicinal importance. The associated factors like rainfall, temperature, relative humidity, soil type and water level data have also been collected. The correlation of types and abundance of plants has been attempted with the characteristic of study areas.

Results and Discussion

The present study has been undertaken across three ecological settings existing within Jodhpur district. These settings represent a hilly area, plain area and irrigated area. The relevant characteristics of each of ecological zone selected for present study and different species of medicinal plants as occurring in different ecological settings of Jodhpur districts are shown in (Table 1 and Table 2) respectively. In all 18 species were studied from all the three settings.

Table 1. Study areas selected for preliminary work and reasoning for their selection

Sl. No.	Areas Surveyed	Geographical Position	Geological Formation	Water Table	Water Salinity & pH. Value
1.	Hilly: Machia Safari Park Jodhpur City Panchayat Samiti: Mandore Tehsil: Jodhpur District Jodhpur.	Away from the source of saline water, i.e. Luni Basin.	'Rhyolite' from basic to top level. Some times presence of patch of clay and sand was observed on the top most level, along with Rhyolite. Rhyolite is not permeable although it shows fractures or joints in between due to "tectonic effects," through which, rain water or fresh water is accumulated underneath.	Medium	No salinity.
2.	Plain Sangariya, Jodhpur District, Tehsil – Jodhpur.	Adjacent to saline water resource, i.e. Luni Basin.	Top layer is formed by alluvium which consists of sand, gravel & silt. Basic layer is formed by granite.	Medium	Brackish water
3.	Irrigated Tinwari and Mathania, Jodhpur District Osian- Tehsil.	Away from saline water resources i.e. Luni Basin.	Top layer is 10-15 ft. Soil followed by sand stone & basic is formed by Rhyolite.	Good	Sweet

Table 2. Distribution of medicinal plants in three ecological zones

S.N.	Dominating Plant Species	Study Areas –1 (Hilly)	Study Areas –2 (Plain)	Study Areas –3 (Irrigated)
	Type of Plants	Herbs & Shrubs	Trees	Trees, Herbs & Shrubs
1.	*Euphorbia caducifolia*	Present	Absent	Absent
2.	*Acacia senegal*	Present	Absent	Absent
3.	*Prosopis cineraria*	Present	Present	Present
4.	*Dicoma tomentosa*	Present	Absent	Absent
5.	*Acacia nilotica*	Present	Present	Present
6.	*Calotropis procera*	Present	Absent	Present
7.	*Aerva persica*	Absent	Absent	Present
8.	*Capparis decidua*	Absent	Present	Present
9.	*Citrullus coldcynthis*	Absent	Absent	Present
10.	*Grewia tenax*	Present	Absent	Absent
11.	*Commiphora wightii*	Present	Absent	Absent
12.	*Aspergus racemosa*	Present	Absent	Absent
13.	*Anogesus pandula*	Present	Absent	Absent
14.	*Pulicarias angustifolia*	Present	Absent	Absent
15.	*Bistella digyna*	Present	Absent	Absent
16.	*Cassia italica*	Absent	Present	Absent
17.	*Crotolaria burhia*	Absent	Present	Present
18.	*Solanum surattense*	Absent	Present	Absent

Table 2 clearly depicts that out of 18 species of medicinal plants, 12 species were present in study area-1 *i.e.* hilly area while in studty area-2 i.e. plain area which is characterized by saline river Luni only 6 plants were present. In the study area 3 characterized by properly irrigated conditions of soil and other factors, out of 18 only 7 plant species were present. The hills in this region were composed of granite, rhyolite, adamellite, quartzite and sand-stone. Rhyolite hills have smooth surface and are covered with scrub community of *Euphorbia caducifolia, Acacia senegal, Prosopis cineraria, Dicoma tomentosa, Acacia nilotica* and *Calotropis procera* as a dominating plants. Other available species present are *Grewia tenax, Commiphora wightii, Asperagus racemosa, Anogeissus pendula, Pulicarias angustifolia* as co-dominating plant species.

In plain areas, top soil is composed of alluvium which contains sand, gravel and silt. The most dominating plant species present were *Prosopis cineraria, Acacia nilotica* and *Salvadora oleoides* while co-dominating species present were *Crotolaria burhia* and Solanum *surattense.* Others have not been recorded.

The third ecotype studied was irrigated area where *Prosopis cineraria, Acacia nilotica, Calotropis procera, Aerva persica, Capparis decidua* and *Citrullus colocynthis* were present.

Present study involves a distinct hypothesis, weather, ecological and edaphic factors associated with occurrence of medicinal plants be developed as determinants or predictors of their occurrence for other areas? A pilot study undertaken to test this hypothesis is presented in this paper. The observation clearly shows that a hilly ecosystem favors occurrence and abundance of medicinal plants in an area. The parameter studied *viz*; such as soil type, topography, ecological factors and human interventions etc. shows a strong role of soil factors for the presence and absence of the particular plant species in an area. The present pilot work needs to be developed as a comprehensive study through which associations can be converted into correlations and ultimately the determinants.

References

1. Bhandari, M.M. 1990. Flora of the Indian Desert.
2. Satyavati, G.V. and Gupta, K.A. 1987. Medicinal plants of India, Vol.1 & 2.

3. Blatter, E.J. and Hallberg, F. 1981-21. The flora of Indian desert (Jodhpur & Jaisalmer).

4. Chopra, R.N; Nayer, S.L and Chopra, I.C. 1956. Glossary of Indian Medicinal plants, Council of Scientific and Industrial Research, New Delhi.

MEDICINAL PLANTS - THEIR GOOD USES AND PLACE IN FORESTRY RESEARCH EXTENSION

Ram Harsh Singh

Vice Chancellor, Rajasthan Ayurved University, Jodhpur

ABSTRACT

Plants have been the eternal source of food and medicine since antiquity in all traditions and cultures. The medicinal plants are considered the most vital component of ecosystem and biodiversity essential for the existence of life besides their economic utility. With the increasing demand and trade potential of medicinal plants, their conservation, cultivation and related R & D activities deserve promotion and as such Forestry Research Extension is on cards.

Introduction

Plants have been the eternal source of food and medicine since antiquity in all traditions and cultures. Prior to the rapid emergence of synthetic chemotherapy in western medicine, the medicinal plants were the sole source of medication. However, in recent years with the growing awareness about the toxicity and side effects of chemotherapeutic agents, the interest in green pharmacy and natural products is fast increasing. India and China have been the leaders in this field because of their sound codified traditional systems of medicine on one hand and rich biodiversity on the other. The present day scenario is global, and the medicinal plants and green pharma products are on high demand world over. It is estimated that the current world market of such products is more than 100 Billion US Dollars which is likely to rise to 5.00 Trillion USD by mid of the present century. The domestic consumption of plant products is also growing fast. There are about 10,000 licensed pharmacies manufacturing ayurvedic drugs.

Because of the above mentioned need-growth, attention has been drawn to cultivate medicinal plants at commercial level. This sector is getting gradually organised. Establishment of Medicinal Plants Board, enforcement of WHO Guidelines on Good Agricultural and Collection Practices (GACP), the new Code of Good Manufacturing Practices (GMP) for Green Pharma Sector and emerging new educational program like Diploma and Degree courses in Medicinal Plants Sector have already exhibited a kind of Medicinal Green Revolution.

At this juncture, the Forestry Research Extension in this sector has outstanding potential. Forest and Agriculture have to tap this newly emerging field of useful activity to promote environment, to generate livelihood, trade and commerce.

Concerted efforts are needed to identify the medicinal trees, shrubs and herbs which could be promoted in different areas with the view to protect environment and to generate now employment and livelihood. There is also a need to workout the economics of each of these plants with market potential which could be propagated among the policy makers, cultivators and the people at large. There is also a need to take help of newer technologies for R & D activities in this field such as Tissue Culture and improved methods of conservation and cultivation of medicinal plants for better quality and yield besides evaluation of their therapeutic value.

A large number of plants are known to possess a wide range of therapeutic applications. About 2000 plants are described in authentic ayurvedic scripts while another 5000 plants are prevalent in traditional folk practices. The researches reported in recent literature including the work of this author, indicates that many important therapeutic indications of different medicinal plants have already been validated with scientific evidence. The most potential therapeutic uses of these plants as scientifically substantiated include anti-stress and adaptogenic activity, anti-inflammatory effect, hypolipidemic effect, cardioprotective effect, immunoenhancer effect, nootropic effect, antilithic effect, anticancer effect and several other such activities. Many of these plant drugs have useful nutraceutical effect and may prove to be good food supplements while several plant products are also used as antioxidants and as antiageing agents. Thus, it would be seen that most of these potential indications are of fundamental significance from the point of view of contemporary health problems and hence plant medicine opens newer vistas of research and development in contemporary health care. Hence forestry research extension in this area will be fruitful.

Thousands of formulations made out of these medicinal plants are in use. Nearly 10,000 licensed pharmacies are working in India on commercial footings in producing such medicines for official use in the medical profession including exports.

Forestry research extension in this area could prove a big boost. The Forestry Research Centres should establish independent units to handle this new area and should develop strategies for sustainable multi-tier cultivation of different medicinal trees, shrubs and herbs suited for different climates and soils with newer strategies of crop diversification. Attempts should be made to strengthen documentation and new technology entrepreneurship.

Biodiversity and Planetary Health

Biodiversity is the most essential need for human survival and planetary health. Biodiversity refers to the variety and variability of organisms and ecosystems which need to be conserved to ensure life and human health.

In the context of Indian Desert, the conservation of biodiversity is essential because of the most fragile ecological processes and life support systems in the desert. There is a need of an integrated conservational strategy for saving of species and meeting the increasing demand of medicinal plants. In this context, forestry has to play the most vital role. The forest policy needs to be designed accordingly.

Sustainable Development and Herbal Farming

Herbal Farming as a component of sustainable development has three-fold advantages. It sustains environment, creates newer avenues of employment and help in promotion of Health of the people in Natural way. The issue is especially significant in the Indian context because of a very sound tradition of ayurvedic medicine where there is a huge consumption of medicinal plants for therapeutic usage involving big commercial potential. It is estimated that the domestic pharma alone amounts to Rs. 20,000 crore beside tremendous scope of exports. Hence herbal farming is an important growing component of sustainable development and livelihood security in India.

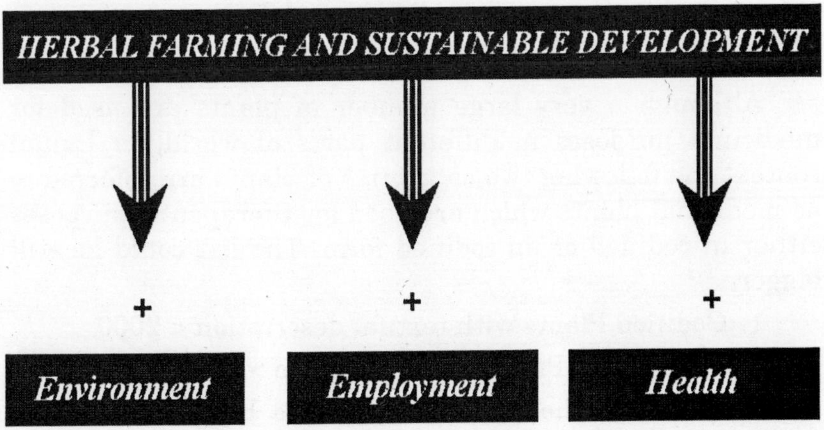

Medicinal Green Revolution

Because of the possible positive impact on planetary health, relatively guaranteed safety, cost effectiveness, pronature activity and universal acceptability besides holistic and nutraceutical effect, the medicinal plants are the most popular mode of medication world over. It is believed that about one third of the world population depends on herbal resources for its health needs. Because of such a growing public demand there has occurred a big boom in world market of herbals and natural products. The following projection would indicate that a medicinal green revolution is taking place:

- Current Global Market: US$ 150 billion (WHO)
- Projection for 2050: US$ 5000 billion (WHO)
- Chinese Foreign Exchange: Rs. 22,000 Crores
- Indian Share: Rs. 500 Crore only
- Chinese Plant Resource: 30%
- Indian Plant Resource: 30%
- Indian System of Medicine: Time tested, trusted and well structured
- Chinese System of Medicine: Not so well trusted and structured

Medicinal Plants in Current Therapeutic Use

Although a very large number of plants are used for medicinal purposes in different parts of world, in Indian context the following two categories of plants are referred to as medicinal plants which are used for therapeutic purposes either in codified or un-codified form. The list could be still bigger.

1. Codified Plants with textual description < 2000
2. Un-codified Plants in Folklore use > 5000

Only few of the above plants have been subjected to systematic and exhaustive Phytochemical and Pharmacological studies upto a logical and conclusive level. The entire range is waiting for intensive research and extension activity. There is a need of extended studies and research on medicinal plants including their conservation, cultivation, value addition, marketing and a range of other R & D activities.

Some leading Phytochemicals isolated from Medicinal Plants and their use

In recent years, the researches conducted on different medicinal plants have lead to the identification of hundreds of new phytochemicals which have been in popular therapeutic use. More and more such studies are warranted to help us to know the specific chemical nature of plants which is of prime need in ascertaining the specific efficacy of the plants and also in their standardization, quality assurance and toxicity monitoring. Few important phytochemicals are enlisted in the following table:

Active Chemical	Medicinal Plants	Therapeutic Use
Reserpine	*Rauwolfia serpentina*	Hypertension
Piperine	*Piper longum*	Bio-enhancer
Digoxine	*Digitalis*	Cardiotonic
Levo Dopa	*Mucuma pruriense*	Parkinsons Disease
Curcumine	*Curcuma longa*	Anti-inflammatory
Vincristicine	*Vinca rosea*	Anti cancer
Kutakin	*Picrohiza kurroa*	Hepatoprotection
Ephedrine	*Ephedra*	Anti-asthmatic
Strychnine	*Nux vomica*	Neuro stimulant
Atropine	*Atropa bellodona*	Anti-spasmodic
	Boerhaavia diffusa	Diuretic

Bacosides	*Bacopa monniera*	Nootropic
Withanolides	*Withania somnifera*	Adaptogen
Glycyrrhizin	*Glycyrhiza glabra*	Anti inflammatory

WHO Alma Ata Declaration 1978 "Health for All"

The Alma Ata declaration of World Health Organization of 1978 overtly admitted that "The target of health for all cannot be achieved without the help of Traditional Systems of Medicine and Local resources". It means that promotion of herbal sector is essential to sustain health care system of different countries including developed countries.

Site-Specific Multi-tier Agro-forestry Models

Promotion of herbal farming and cultivation of medicinal plants is the prime need in the present context. Till recently the herbal drug industry has been dependent on wild natural resources of herbs including forests. But because of heavy increase in demand of these herbs in industrial sector leading to exhaustion of wild natural sources, it has become absolutely necessary to conserve these plants at source and to generate new artificial agricultural resources by strategic cultivation of medicinal plants. This sector is new for India and hence there is a need to develop appropriate Agro-technology and Agro-forestry models for this purpose suitable for different plants cultivable in different soils and climates. There is also a need to identify the range of plants which may be cost-effective to cultivate and which may become a source of better income for the farmers. The main criteria for selection of plants for cultivation should be (1) Therapeutic utility, (2) Commercial potential and market demand, (3) Eco-friendly and (4) Easy cultivation. Technology of diversified farming and multi-tier agriculture system may prove economically viable and cost-effective method of cultivation for medicinal plants. The following schema exhibits certain site specific multi-tier agro-forestry models.

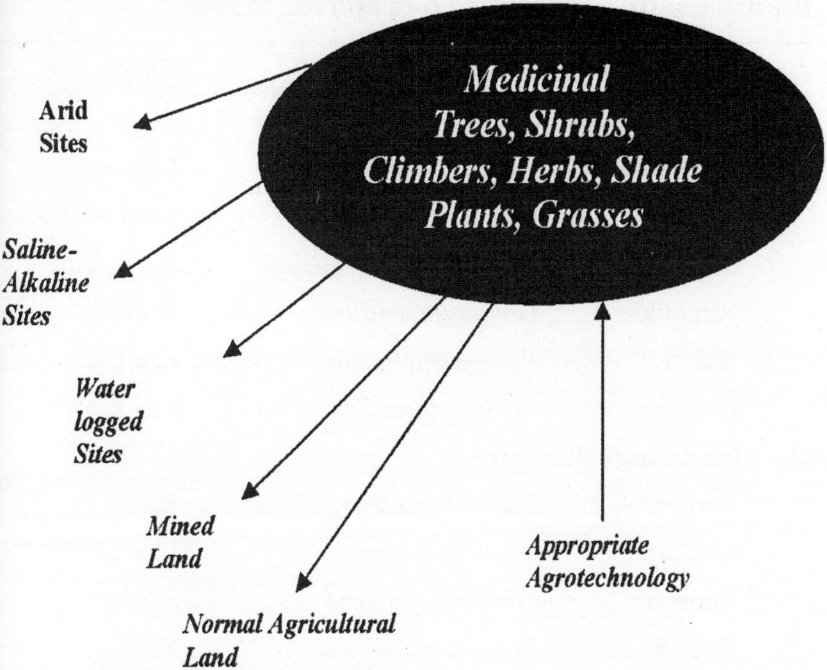

Some Popular 3-Tier MTAS

A common MTAS model proposed by some experts and also in use by certain cultivators is to grow simultaneously three kinds of medicinal plants on the same piece of land. This involves planting some base plants which may provide limited shade for shade loving plants planted below the base plant, besides climbing support for certain appropriate climbers. If suitably planned, such models may prove cost-effective. Thus, three tier MTAS models may be designed utilizing the following kinds of Medicinal Plants:

1. The Base Plants which may provide shade for shade loving plants and support for Medicinal Climbers.
2. Medicinal Climber plants as co-crop.
3. Medicinal shade loving plants for co-farming.

Table showing Some Good Base Plants suitable for Western Rajasthan and medicinal climbers and shade plants useful for use as co-crops for MTAS

	Name	Botanical Name	Usable Part
1.	**Base Plants:**		
	Ratanjot	*Jatropha curcas*	Leaves
	Arand	*Ricinus communis*	Seed oil, Root
	Shigrue	*Moringa pterygosperma*	Bark, fruit
	Nimbuka	*Lemonum acidum*	Fruit
	Anara	*Punica granatum*	Fruit
	Karmarda	*Carissa corandus*	Fruit
II.	**Medicinal Climbers**		
	Hadjoda	*Cissues quadrangularis*	Stem
	Kalihari	*Gloriosa superba*	Leaves
	Kaunch	*Mucuma pruriens*	Seeds
	Pippali	*Piper longum*	Fruit
	Karela	*Momordica charautia*	Fruit
	Guduchi	*Tinospora cordifolia*	Stem
III.	**Shade Plants**		
	Sweta Musli	*Clotophytum boribiliamum*	Tuber
	Shatawari	*Aspargus racemosus*	Tuber
	Haridra	*Curcuma longa* *Acoras calamus*	Tuber

WHO Guidelines on Good Agricultural and Collection Practices (GACP)

With the rapid increase in use of herbal medicines world over during last two decades, a number of reports and experiences have revealed lack of quality and genuineness of medicinal plant materials in use which is becoming a matter of concern. This situation prompted World Health Organization to issue (WHO, 2003)[11] Guidelines on Good Agricultural and Collection Practices (GACP) for medicinal plants with specific guidelines on following aspects.

1. Identification and authentication of cultivated medicinal plants *viz.*, selection of plants, botanical identification and herbarium etc.
2. Seeds and other propagation materials - their availability and standardization.
3. Cultivation *viz.*, site selection, ecological, environmental and social impact, climate, soil, irrigation and drainage, plant maintenance and protection.
4. Standard harvesting procedures and trained personnel for harvesting.
5. Good collection practices - selection, standard collection procedures and technical planning.

Economics of Medicinal Plants

Till recently, it has been in practice to harvest and collect medicinal plants, growing wild in forests, as well as, on wastelands, free of cost with expenditure on labour and transportation. With the growing demand and high consumption, and due to impending danger for the popular medicinal plant species likely to be extinct, it is now imperative to encourage cultivation of high demand herbs through organized farming. Gradually, herbal farming is now emerging as diversified farming for rural livelihood security.

In view of this prospect, the Medicinal Plants Board has already done pilot exercise to work out the economics of many medicinal plants. List of some such plants indicating significant profitability to the farmers is given in table.

Table showing the Economics of some selected medicinal plants

Species	Duration	Input cost (Rs.)	Output (Rs.)	Yield
Amla	After 6 years	15,000	50,000	10 t fruit
Ashok	10 years	5,500	20,000	- bark
Ashwagandha	1 year	6,000	19,500	3 q dry root
Bhumyamalaki	1 year	2,000	4,000	5 q herbage
Brahmi	1 year	15,000	50,000	20 q herbage
Guduchi	1 year	2,500	6,000	4 q dry stem
Guggulu	After 8 years	100,000	200,000	8 q gum resin
Konch	1 year	4,000	9,000	5 q seed
Kumari	1 year	8,000	12,000	50 q fresh leaf
Mulethi	3 year	35,000	180,000	30 q dry root

Source : Rajasthan State Medicinal Plant Board, Department of Ayurveda, Govt. of Rajasthan, 2002

Therapeutic Value of Some Plants[1-10]

Medicinal Plants have been described and have also been found to clinically possess a wide range of therapeutic efficacy. However, the recent researches conducted on medicinal plants in recent past have yielded evidence and leads for the following categories of therapeutic effects. It is presumed that the following are the priority areas of therapeutic testing of medicinal plants.

- Antistress and neuro-active drugs[7]
- Immunomodulators and adaptogens
- Anti-inflammatory drugs[9]
- Cardio-protectives and hypolipidemics[9]
- Hepato-protective remedies[9]
- Nutraceuticals and antioxidents
- Anti-aging, anti diabetic & anti-cancer drugs[2, 10]

References

1. Archana R. and Namasivayam, A. 1999. Antistress Effect of *Withania somnifera*. J. *Ethnopharmacol* 64 91-93.
2. Dhar, M.L.; Dhar, M.M.; Dhawan, B.N.; Mehrotra, B.N. and Roy, C. 1968. Screening of Medicinal Plants for biological activity-1 *Ind. Jour Expt. Biology*, 6: 232-247.
3. Mangain, P. and Singh, R.H. 2005. Cardioprotection. In Advances in Ayurvedic Medicine Vol 3 PP 70-169. Choukhamba Publications, Varanasi.
4. Sharma, P.V. 2000. *Dravyaguna Vigyan* Vol. 1-5 pub. Choukhamba publications, Varanasi.
5. Singh, R.H. and Sharma, G.G. 2005. Arid Shrubs for human health. In shrubs of Indian Arid zone (ed.) Pratap Narain *et al.*, Arid Agro-ecosystem Directorate CAZRI, Jodhpur PP 39-44.
6. Singh, R.H. 2006. Prospects of cultivation of Medicinal Plants in Arid Region. National Symposium on Livelihood Security and diversified farming systems in Arid Region, Jan 14-16, 2006. Arid Zone Research Association of India, Jodhpur PP 18-22.
7. Singh, N. 1987. Antistress effect of Medicinal Plants. *Ann. National Acad Ind*. Med. Vol 1 No.1.
8. Singh, R.H. 2002. The Holistic Principles of Ayurvedic Medicines. Speicial monographs published by Choukhamba publications, New Delhi.
9. Singh, R.H. and Associates. 2005. Advances in Ayurvedic Medicines Choukhamba publications, Varanasi.
10. Smit, H.F.; Woerdenbag, H.J.; Singh, R.H. 1995. Ayurvedic Herbal drugs with possible cytotoxic activity. *Jour. Ethnopharmacol* Vol. 47: 75-84.
11. WHO. 2003. Guidelines on Good Agricultural and Collection Practices (GACP) for Medicinal Plants, World Health Organization, New Delhi.

FORESTRY EXTENSION IN HOT ARID TRACT: WHY ARE WE NOT SO SUCCESSFUL

M.A. Khan, J.C. Tewari and Pratap Narain

Central Arid Zone Research Institute,
Jodhpur – 342 003

ABSTRACT

The philosophy of extension in the original and wider sense is the knowledge of things and their causes, both theoretical and practical. In arid tract, majority of forest lands are in the form of shifting and partially stabilized sand dunes. Agriculture extension has sound roots in the country, while forestry extension is still at its infancy. The paper describes the approaches to develop extension system in forestry sector. Technology dissemination in forestry sector is still very poor and requires initiatives for better forestry/agroforestry extension. The stakeholders need clear-cut information and inputs, which at the moment is lacking. Transparent forestry extension programmes can change the production system scenario in general and arid regions in particular.

Introduction

If we talk in general regarding extension programme in India, of course, which were essentially agriculture-livestock based, as compared to pre-independence efforts post independence efforts have seen a tremendous expansion, which has no parallel in the history of the developing countries. The philosophy of extension in the original and wider sense is the knowledge of the things and their causes, both theoretical and practical. The philosophy of extension has been described and interpreted in different ways by different authors and a clear picture can not be drawn due to very complexity of its nature (Dhama and Bhatnagar 1993). Kesely and Heavne (1963) described that the philosophy of extension work is based on importance of the individual in the promotion of progress of rural people and for the nation.

Rewards in extension may be in the form of material, money, symbolic, public, recognition, etc. They are effective devices to motivate the rural people, as well, as extension workers. True motivation of stakeholders is the key of development of particular production system. If we see the extension strategies in India, we find that only agriculture and livestock production system were targeted most and accordingly the achievements are tremendous. However, the forestry sector, a very high production system, remained neglected as far as extension programmes were concerned. Historic evidence indicates that forest/woody resources were adequate to meet villager's requirements upto first half of the 19th century. In fact, before the Forest Conservation Act 1980, villagers had traditional responsibilities for the management of forest or forest lands at and around villages (Gadgil, 1989). The system functioned smoothly without any damage to forest wealth for many generations, whilst the population was low. However, when population increased and the commercial interests of the state resulted in declaring much of the forest lands as reserved forests, the rights of villagers to use them were withdrawn through out the country and conflict for resources between villagers and state intensified. Village communities were completely alienated from management of the forest and, as their needs grew, this was accompanied by their ruthless exploitation and consequent degradation of tree cover.

The Situation in Hot Arid Tract

The hot arid zone of India lies between 21° and 29° North Latitude, and 70° and 76° East Longitude, involving seven states, *viz.*, Rajasthan Gujarat, Punjab, Haryana, Andhra Pradesh, Karnataka and Maharastra. In Rajasthan state, hot arid areas are spread over the western and north-western parts, which accounts for 61% (208751 km^2) of the total Indian hot arid zone. Thus, the region forms principal hot arid zone of the country and is also referred as the *Thar* desert. The mixed crop-livestock farming, mixed livestock-crop farming and livestock farming form the spectrum of economic activities from settled agriculture (in true, sense, settled agroforestry) to nomadism (particularly pastoralism) (Tewari, *et al.,* 2003). Growing arable crops in association with multipurpose tree or shrub species (MPTS) is an age old practice of the region, commonly referred as traditional agroforestry. These traditional agroforestry systems are unique form of combined productive-protective system practiced as private production system. This form of wide spread private land forestry has its own advantages, though some problems too. Farmers are managing such system since time immemorial, though no-where such agroforests are mentioned in land and revenue records. In fact, forest cover in arid western and north-western Rajasthan is less than 1% and only these agroforestry systems are providing some amount of tree cover.

The majority of forest land is in the form of shifting or partially stabilized sand dunes in arid tract. More-over, plantations on IGNP canal form large chunk of afforested areas. The farming systems of entire hot arid tract represents an agriculture-tree/shrub-livestock-man continuum and as such the importance of trees is many fold higher for life support than the humid, semi-humid, temperate and other agro-eco-climatic zones capable of supporting good woody vegetation. Unfortunately, till date the attempts on forestry or agroforestry extension are almost negligible. This paper attempts to discuss the role of extension system in forestry/agroforestry development, and the reasons behind the very slow growth in forestry extension sector even after 58 years of independence particularly in arid tract.

Agricultural Extension

Food is essential for human beings. In general, entire animal kingdom sustains on food. After independence country felt major shortage of food grain and to restore adequate food supply there were only two alternatives, either import the grain or increase their production to feed ever increasing human population. First decade after independence some how adequate food grain supply was maintained through import from developed countries mainly USA, Australia, etc. Early 1960s to mid 1960s saw unprecedented food grain shortage resulting in famine like situation in many parts of the country. However, as a blessing in disguise the collective efforts of agricultural scientists in the leadership of Dr. M.S. Swaminathan and Dr. Norman Borloug; sound political will of late Prime Minister Mrs. Indira Gandhi and her agriculture minister Bharat Ratna Mr. C. Subramanium turned the table and country marched forward in the path of green revolution. This was followed by white revolution and blue revolution, and consequently country attained a status of self reliant in food and related commodities. During the process of green revolution, with passage of time cohesive agriculture extension sectors have been developed in different states of the country and in some states they became so much strengthen that with their involvement economies of thousands of villages turned from 'begging bowl' to "exportable surplus".

When the question of forests and forestry comes with exception of 'Van Mahotsava' a one day yearly ritual, there was nothing upto late 1970s as far as forestry extension programmes were concerned. We never look forestry beyond a production system for residual lands. Though overall importance of forestry in national economy is not too much, however, services of forests are very important, whether they have a formal market (such as for grazing, eco-tourism, fuel wood, minor forest produce, recreation, etc.) or are largely intangible (such as land conservation value, cultural and spiritual values or biodiversity conservation).

In hot arid tract, where agroforestry is practiced since time immemorial, there was rapid decline in superior forest/tree biomass in last quarter of 20th century. Development of sustainable management is key component of an overall trend toward globalization in the forestry sector (FAO, 1988). One of the most important frontiers of people/forest - tree interactions concern farmers and other outside forests that depends on forest/tree produce for various needs and supplementary opportunities. For example, 70% rural population of arid and semi-arid tropics of India, depends on fuelwood for their fire energy needs. In arid tract of India still about 25% of fodder for livestock is obtained from trees/shrubs (Khan *et al.* 2006). These few examples clearly indicate how much forestry extension is important for sustainable supply of fuel and fodder for rural folk.

Forestry Extension: Challenges and Opportunities

The first forest policy framed by Britishers were absolutely devoted to commercial interest of the State and restricted the legal rights of villagers to fuel, fodder and minor forest produce harvesting. The first forest policy of 1952 of independent India tried to maintain balance between rights of state and villagers' legal rights. However, the major exploitation of forest wealth occurred between early 1950s and late 1980s. The forest policy of 1988 first time considered forest conservation as most important issue and since then great strides have been made on conservation oriented forestry.

Forests are re-newable source of energy and as such with appropriate conservation strategies, the harnessing of forest produce is also essential for benefit of stakeholders. The forests are directly or indirectly related to agricultural production. According to Singh and Singh (1992) in the western Himalayan region, each hectare of cultivation required 20 ha of forest to meet villager's need of fodder and 12 ha of forest to meet their firewood needs. In agroforestry systems, trees also provide considerable amount of energy for smooth functioning of the agro-eco-systems.

In spite of all these facts related to forest and trees, and their role in agro-ecosystem functioning, the forest education was negligible in the country except giving managerial trainings to forest officials. Apart from Forest Research Institute, Dehradun there were only few state forest Institutes. All these institutions were primarily devoted to timber forestry. Mean while many schemes like social forestry, farm forestry and urban forestry came time to time, but unfortunately the success of such schemes remained much below than the expected level.

With creation of Indian Council for Forestry Research and Education (ICFRE) in late 1980s many national level subject matter specialist institution of forestry were established. This development gave some driving force to forestry extension education system. Side by side many agricultural and traditional universities also established forestry departments and/or faculties.

Still, the forestry extension systems in various states of the country are in their infancy. It is a challenging task to make such systems strong and more stakeholders friendly. Rural folk are primary user of forest produce but, despite their dependence on forest/tree produce, they have no acknowledgement for the same. This is the point which requires utmost priority in forestry extension sector. An appropriate extension strategy is needed to educate rural folk regarding importance of forests or trees. Forest department of each state should have an independent Forest Extension Division. This division be made responsible for educating people regarding importance of forest and trees. Private forestry activities, especially by rural folk should be encouraged. For such private forestry activities, farmers should be given proper training from the point of nursery raising to out-planting seedlings on fields and their proper management for harvesting maximum returns. Returns can be any thing like tree leaf fodder, fuel wood, gum and resins, timber, etc.

Private forestry sector need immediate attention on the part of policy makers, forest managers, scientists, NGOs, etc. Such forestry activities should be need based and location specific. Market infra-structure facility development in small scale is of prime importance and therefore, should be given very high priority. Such interventions can open doors for private land forestry.

Through out the world, forests are being exploited ruthlessly. If, some how, private forestry could be organized successfully, it will definitely release pressure from natural forests and will ultimately play great role in ecosystem and biodiversity conservation. If the forest extension agencies are strong, they can forcefully deliver the message among mandarins. Once the people will realize the importance of forests/trees, and their services to mankind, the whole scenario can be transformed. The role of extension in agriculture development in the country is living example of quantum of success, can be achieved through innovative extension ideas. For development of appropriate forestry inter-system, following points should be taken into consideration:

- Integration for forests/trees into landuse plan,
- Increasing knowledge and awareness on all aspects of trees/forests,
- Promoting policy initiative favouring sustainable forest/tree management and use,
- Clarifying tree tenure among stakeholders,
- Developing management strategies based on local needs and knowledge, and
- Ensuring better assessment and equitable sharing of costs and benefits of forest/tree management, and use among all stakeholders.

Actions on these points will require collaborative efforts by forest managers, research institutes, NGOs, developmental agencies, local communities and policy planners.

Technology Dissemination: Most Powerful Tool for Adoption of Forestry Programmes

Once the technology is perfected, it requires appropriate dissemination with such strategic features that stakeholders themselves started to adapt with whatever inputs they are getting. The best way to disseminate any forestry based technology should provide all the information without any hidden aspect. There should be complete transparency on the part of technology dissemination process which includes (i) information, (ii) consultancy, (iii) inputs supply, (iv) technical service, (v) marketing service, and (vi) other services such as insurance, infrastructure, etc.

Apart from increasing the growth and production of tree based system (forestry, agroforestry, social forestry, farm forestry, etc.) and standard of living of stakeholders, a strong intra and inter-personal relationships with various government bodies and other agencies must be developed with stakeholders. This reiterates the fact that in any production system strong personal relationship between developmental agencies and the community have far greater impact than a supplier-receiver relationships in which relationship dies off as soon as the supplies are withdrawn. With the active support of the organization, the developmental agencies too found it easier to implement forestry based programmes resulting in a mutually beneficial relationship. This, forestry extension in rural setting of the country requires both government initiatives and institutionalization of private extension.

Epilogue

The local communities and even commercial enterprises increasingly rely on forest/tree based wide range of products and services, given the extensive and accessible nature of this resource base (wood resources). The trees or forests on private lands are threatened by high extraction pressures and repeated disturbances with poor regeneration, leading to progressive degradation. Till date, there is no clear policy even regarding well established traditional agroforestry systems. In such a situation, talks of other community related forest programmes have no meanings. Why we could not achieve desired results in community forestry programmes reflects the weakness of our forestry extension sector. The need of hour is to frame very clear and precise forestry extension policy so that a movement could be generated for community oriented forestry programme.

References

Dhama, O.P. and Bhatnagar, O.P. 1993. *Education and Communication for Development* (sixth printing). Oxford & IPH Publishing Co. Pvt. Ltd. 734 p.

FAO. 1988. Training for Agriculture and Rural Development, F.A.O. United Nation, Rome.

Gadgil, M. 1989. Deforestration: Problems and Prospects. *Foundation Day Lecture* (*mimeo*). Society for Promotion of Westland Development, New Delhi.

Khan, M.A., Tewari, J.C., Singh Raj and Narain Pratap 2006. Structure, Production attributes and management strategies in a traditional extensive agroforestry system in an arid region watershed of India. *Forests, Trees, Live, Livelihoods*, Vol. 16 (in Press).

Kesely, L.D. and Heavne, C.D. 1963. *Cooperative Extension Work.* Cornell University Press, New York.

Singh, J.S. and Singh, S.P. 1992. *Forests of Himalaya: Structure, Functioning and Impact of Mean,* Gynodaya Prakashan, Naini Tal, India.

Tewari, J.C., Tripathi, D., Narain Pratap and Singh, S.P. 2003. A study of structure, energy fluxes and emerging trends in traditional Central Himalayan Agroforestry System. *Forests, Trees and Livelihoods* 13: 17-38.

STRATEGIES OF TREE INTEGRATION TO CONTROL LAND DEGRADATION AND TO INCREASE AGRICULTURAL PRODUCTIVITY OF DRY LANDS

G. Singh
Division of Forest Ecology, Arid Forest Research Institute,
New Pali Road, Jodhpur-342 005, India
E-mail: gsingh@afri.res.in

ABSTRACT

Growing trees in association with agricultural crops, grasses or medicinal herbs enhance overall productivity of the cropping system in addition to the environment amelioration, carbon sequestration and phyto-sanitary benefits. However, imbalances in the sharing of resources like water, nutrients and light between tree and associated crops, results in either an increase or decrease in production depending upon types of interaction i.e., facilitative or competitive. Facilitative effects enhance productivity whereas competitive effects reduce the productivity of associated crops. Though tree provides many benefits like product diversification and enhancing carbon stock but long gestation period and competitiveness with associated crops make the farmers more reluctant in tree integration on their farmland. Thus, integration of short-term benefits of diversified crops, compensatory benefits on crop reduction and monitory benefits in increasing carbon stocks may be helpful in tree integration on farmlands that improve production and control land degradation. Many technologies are in way like agroforestry, windbreaks, shelterbelt, woodlot development, plantation of multipurpose trees, their silvicultural management and generation of alternative livelihoods to benefit the local people. Some strategies on people mobilization for tree integration and afforestation may be use of mass education about tree benefits, compensatory benefits on crop loss and carbon credit generated by stockholders. The research results on (i) effect of tree density on crop yield, which increase when density of tree is optimum, though it varies with tree size/age; (ii) effect of crop sequence on overall productivity; (iii) species suitability in agroforestry; and (iv) species suitability in sand dune stabilization, interim benefits and the benefit of carbon sequestration has been described as the tools for tree integration.

1. Introduction

About 40% (i.e., 61 million km^2) of the world's land area is under arid and semi-arid region (NAPCD, 2001). Land management practices like afforesting agricultural land, reduced tillage, improvement in soil water status, organic mulching and re-vegetation of sand dunes or saline alkaline lands etc. may increase biomass and organic carbon content in soils (Havelin, *et al.*, 1990) in addition to environmental and aesthetic benefits. However, low and erratic rainfall, intense solar radiation, high wind velocity, recurring drought and famines are common features of the Indian arid zone that results in high mortality, low productivity and low carbon gains per unit area.

Integrating trees in agricultural farms and afforesting stress sites or community lands may be a useful technique to increase biomass, as well as, soil C status as trees affects C dynamics directly or indirectly (Dixon, 1995). Trees and the associated vegetation improve soil productivity through ecological and physico-chemical changes that depend upon the quantity and quality of litter reaching the soil surface, the rate of litter decomposition and nutrient release (Meetenmeyer and Berg, 1986; Bockheim, *et al.*, 1991; Stump and Binkley, 1993; Sharma, *et al.*, 1997). Further, trees provide an assurance to the farmers towards agricultural production in normal rainfall year, while in famines and drought years they provide top feed for livestock. In addition, they conserve carbon, which otherwise goes to atmosphere due to deforestation in absence of agroforestry products (Sanchez and Benites, 1987; Brown *et al.*, 1993). Introduction of surface vegetation in association with above canopy trees not only improve soil conditions but also increase produce of economic importance, sequester carbon and reduce land degradation by wind erosion and sand drift. Despite various benefits of trees like addition of litter, maintenance of higher soil moisture regime, reduced surface soil temperature, proliferated root system and enhanced biological activities and decreased risk of soil erosion (Blevines and Frye, 1993; Lal, 1989; Singh and Rathod, 2002) people are reluctant in integrating trees on their farmlands or growing trees because of low returns, absence of alternative livelihoods and absence of promotional benefits. This requires rethinking and framing suitable strategies for benefiting all i.e., local community, stack-holder and country as a whole.

This manuscript describes the importance of growing trees in association with agricultural crops or grasses in the form of economic benefits from the produce, highlighting the compensatory benefits for increasing carbon stocks in the dry ecosystem and some suggestions on effective people participation in afforestation programmes.

2. Methodology and approaches

2.1 Site conditions

The north western region of India covering Thar Desert comprises various topographical condition including sand dunes, inter-dune plains, rock outcrops, saline flats etc. The region is characterized by extremes of temperature (rising up to 49°C in summer and dropping to minimum of 0°C in winters), low and erratic rainfall, and strong winds. Rainfall was ill distributed resulting in severe moisture deficit conditions. Most of the soil of the region is loamy sand with low in organic matter, available nitrogen and available phosphorus. Soil pH is high.

2.2 Technical strategies

Afforestation progammes should involve agroforestry system involving grasses, agriculture crop, medicinal plants and livestock as the alternative strategies to improve productivity of the land.

Intensification of production systems

These activities will target to increase forest cover/production, crop production and livestock production. They will be aimed at encouraging the stakeholders to increase soil productivity so as to reduce the pressures on natural ecosystems and also to reduce excessive vegetation removal, over-exploitation and land degradation.

Tree integration in agriculture land

Prosopis cineraria is the most preferred species integrated under agroforestry. It is known for improvement in soil and the associated crop productivity. Some other species *viz.*, *Tecomella undualta, Hardwickia binata, Colophospermum mopane, Emblica officinalis* have also been integrated in agricultural systems to diversify the farm income and sustain the livelihood in the area (Gupta, *et al.*, 1998; Singh, *et al.*, 2004).

Sand dune stabilization using surface vegetation

In sand dune stabilization, seedlings of tree species like *Acacia tortilis*, *Prosopis juliflora* and *Calligonum polygonoides* can be planted. *Cassia angustifolia* could be sown to develop under canopy vegetation and generating interim income. Live saving irrigation could be done four times in first year (@ 30 liters per irrigation in September, November, February and April months provided with sand mulching) and three times in second year (November, February and April months) (Singh and Rathod, 2002).

2.3 Income diversification

Beekeeping

Honey production will simultaneously generate sustainable income and protect the community or forest areas where this activity is conducted. Honey production while improving the local diet will also generate sustainable income through sales. The only obstacle to bee keeping is the lack of access to initial micro-credit for the population, so the isolated village is located at the forest periphery. The durability of this activity resides in its economic and financial profitability. For bee keeping, the initial funds required are for the purchase of a protective suit and chemicals to put the bees to sleep for 35 minutes. The honey production season goes from January–August including agricultural, as well as, forestry areas.

Poultry farming

Poultry farming is another source of income generation for the peoples of the dry areas. It not only generates sustainable income, improves local diet but also helps in reducing the population of small ruminants. Durability of this activity depends upon its economic and financial profitability. The main obstacle to this technique is the micro-credit for the population of the isolated villages located in the remote areas.

Hay harvesting

As a consequence of village land degradation, straw to build or repair huts in the villages has become scarce, therefore, expensive commodity. Community land and forest areas are the only place where it is still available. Despite explicit prohibition on grass harvesting in the forest areas and the dangers of confronting with forestry officials, the population indulge in this activity either for their own use or to sell the straw, which is in high demand during scarcity. Promotion of grass growing in community land or forest areas and involving participatory mechanism of its harvesting will not only protect the ecosystems against fire hazard but also generate substantial sustainable income to the people living in the vicinity of the plantation or forest.

2.4 Carbon credits and local benefits

The Kyoto Protocol is the first step towards stabilizing global emissions of carbon dioxide. Tree accumulate substantial amount of carbon in the form of biomass and helps in carbon accumulation in labile and non-labile carbon in soil. Thus, growing tree involve global benefits of carbon sequestration and money involved in carbon trading could very much be shared with the tree growers or farmers. For example, a farmer in the agriculture belt plant trees to arrest erosion/ stabilize dune, salinity and soil degradation, but he can not afford to plant trees usually for not earning money for many years, if ever. Then, along comes a carbon broker. He offers to pay money up-front to plant trees. In return, he wants a credit for the carbon such trees will store. Thus, a farmer can plant the trees and pocket the cash. As more and more farmers switch to trees, the agriculture belt becomes more like a belt of living carbon. In reality, this is the dream of some Australian farmers, foresters and entrepreneurs. They believe that a system of carbon emissions trading that allows the buying and selling of carbon 'credits' will simultaneously help prevent global warming and promote the planting of trees on degraded land.

Compensatory benefits of tree integration

Despite having multiple benefits, slow growth and small returns from indigenous species is sufficient to discourage the people to involve these species in agriculture land. Further, sometimes tree reduce production of agriculture crop growing in vicinity of tree. Thus, for promotion of tree integration, some compensation may be given to the farmer in lieu of crop reduction till the farmers get returns from the growing trees.

3. Some research results for production enhancing

3.1 Productivity under crop sequences

Six year old plantations of *Prosopis cineraria* and *Tecomella undulata* intercropped under six different crop sequences like no crop (C1), Mungbean-Mungbean (C2), Mothbean-Mothbean (C3), Clusterbean-Clusterbean (C4), Pearlmillet-Pearlmillet (C5) and Mungbean-Pearlmillet sequences (C6) for 6 year indicated that *P. cineraria* exerts relatively less competition with agricultural crops. In 1996, Mungbean yielded 222 kg ha^{-1} when sown with *P. cineraria* and 188 kg ha^{-1} with *T. undulata*. The yield of cluster bean was 106 kg ha^{-1} with *P. cineraria* and 77 kg ha^{-1} with *T. undulata* and that of Mothbean was 52 kg ha^{-1} with *T. undulata* and 18 kg ha^{-1} with *P. cineraria*. Straw yield was highest in Mungbean-Pearlmillet rotation (1092 kg ha^{-1} with *P. cineraria* and 1962 kg ha^{-1} with *T. undulata*). Legume crops were more suitable than pearl millet, which extract more water. Crop production was less near tree and increased with distance from tree base. Biomass production from tree was about 25-28% higher under agroforestry than in the control plot where only trees were growing. Biomass yield of Khejri was highest in the treatments in which Mothbean-Mothbean (31.93 tonne ha^{-1}) and Pearlmillet-Pearlmillet (31.93 tonne ha^{-1}) taken as the agriculture crop. Biomass of Rohida was highest (30.26 tone ha^{-1}) when intercropped with Mungbean-Mungbean crop sequence, which was almost double the biomass yield as compared to pure stand (16.65 kg ha^{-1}).

3.2 Productivity under different stand density

An experiment started in July 1991 by thinning one-year old plantation of *Prosopis cineraria* and *Tecomella undulata* to maintain tree densities of 1666 tree ha^{-1} at 2 m x 3 m spacing, 833 trees ha^{-1} at 4 m x 3 m spacing and 417 trees ha^{-1} at 4 x 6 m spacing (Singh, *et al.*, 2005). Trees were intercropped till 1994 when crop production was highest in 417 tree ha^{-1} plot. The plantation was further thinned in June 1995 and June 2002 by removing trees alternatively to reduce competition with the associated agricultural crops and to obtain tree spacing of 4 m x 6 m (D_1), 4 m x 9 m (D_2) and 8 m x 6 m (D_3). The intercrops tested were *Vigna radiata* (Mungbean), *Cymopsis tetragonoloba* (Clusterbean) and *Pennisetum glaucum* (Pearlmillet). Crops were sown and harvested in rainy season (July to October) and under rainfed condition. Results indicated that optimum density based on the agricultural production and resource availability declines with advancing tree age. In 1991 and 1992 (at two and three year of age) optimum density was 833-tree ha^{-1} in which grain yield was highest (350 kg ha^{-1} mungbean in 1992 as compared to 255 kg ha^{-1} in the sole agriculture crop). In 1993, the optimum density declined to 417 trees ha^{-1} with fodder yield of pearl millet as 1772 kg ha^{-1} whereas production of cluster bean in 1994 was 2571 kg ha^{-1}. Study indicated that a density higher than the optimum resulted in reduction of crop yield and tree growth. In 1996, production of pearl millet straw yield was highest at 278 tree ha^{-1} (1829 kg ha^{-1} with Khejri).

At 11 years of age, optimum density observed was 208 SPH. *Prosopis cineraria* provided utilizable biomass of 19.96 tonnes ha^{-1} including leaf fodder (0.85 tonnes ha^{-1}) with *Vigna radiata* grain yield of 193 kg ha^{-1} at the density of 208 SPH, Similarly, in the year 2001 *Penneisetum glaucum* yielded maximum grain (158 kg ha^{-1}) at the same density. Trend in crop yield *Tecomella undulata* plots was similar to that in *P. cineraria* plot but the yield was less. Therefore, 208 trees of *P. cineraria* can be grown per hectare up to the age of 12 year without significant reduction in crop yield. In addition, dry biomass of about 19.0 tonne can be obtained from the tree from the same area. Conclusively, yield of agricultural crop increased when tree density is appropriate (i.e., optimum tree density), which varied with tree size/age because of competition for soil resources. Production of agriculture crop was highest at optimum tree densities of 278 trees ha^{-1} (4 m x 9 m) at 6 and 7 years, 208 trees ha^{-1} (8 m x 6 m) at 10 year and less than 208 trees ha^{-1} at 11 years of age.

Table 1. Agricultural production (kg ha^{-1}) at varying densities of *P. cineraria* and *T. undulata* trees.

Tree Density	1995		1996		2000		2001	
Grain Yield	PC	TU	PC	TU	PC	TU	PC	TU
417	202	185	-	-	153	123	133	122
278	260	198	-	-	162	140	141	131
208	252	215	-	-	193	153	158	136
Control	240	240	-	-	180	180	165	165
Straw Yield								
417	1857	1223	764	1203	1666	1157	4330	3827
278	2225	1335	1829	1550	1699	1423	4430	4100
208	2050	1208	1667	1228	1803	1510	5420	4560
Control	2020	2020	1230	1230	1720	1720	5470	5470

PC = Prosopis cineraria and TU = Tecomella undulata.

Fodder and fruit productivity under Agroforestry

Emblica officinalis (Aonla), *Hardwickia binata* (Anjan) and *Colophospermum mopane* seedlings were planted in July 1994 at 5 m x 5 m spacing with two plots of (i) trees intercropped under rotational cropping and (ii) trees intercropped with *Vigna radiata* only. In addition to these, a control plot of agricultural crops was also laid. Agricultural crops were *Vigna radiata, Cymopsis tetragonoloba, Penisetum glaucum* and *Sesamum indicum*. Results indicated that soil water status and growth of tree seedlings were higher under rotational cropping system. There was increasing trend in soil organic matter pool under all the three species after forth year and the increase is more under *E. officinalis*. Fodder production started after 5 year and *C. mopane* produced 160 kg dry mass ha^{-1} (400 plant ha^{-1}) with 27% leaf dry biomass whereas *H. binata* produced 122 kg ha^{-1} dry biomass with 30% leaf contribution. Levels of PO_4-P, NO_3-N and NH_4-N increased during summer. However, the increase in the concentration of some nutrients was also observed in winter months, which may be attributed to either plant senescence or change in the soil biological processes. Higher availability of NH_4-N in soil under *E. officinalis* indicated relatively lower rate of nitrogen mineralization than under the legume species *H. binata* and *C. mopane* soil. Approximately two to three fold higher concentration of NH_4-N was observed during monsoon (cropping period). Yield of agricultural crops did not differ significantly as compared to the control plot. Though the yield was higher in the above average rainfall year. Agricultural production was higher under agroforestry than in the sole crop plot during low rainfall year. But the trend was found reverse when rainfall was above average. Crop production was comparatively low in *C. mopane* plots than the *E. officinalis* and *H. binata* plots. The study suggests that integration of tree are more beneficial during low rainfall when crop production fails under sole cropping system. *C. mopane* is more suitable to rehabilitate degraded land and can be grown under site preparation and direct seeding.

Sand dune and sand dune stabilization

Tree planted in September 1996 indicated higher height of *A. tortilis* than that of the other species but the surface spreading (crown diameter) of *P. juliflora* and *C. polygonoides* was considerably higher, which is indicative of their quicker role in control of sand drift. Surface spreading crown of these species affected wind speed under the canopy and was beneficial in control of sand movement and growth of surface vegetation during harsh condition. Low soil water content in *A. tortilis* and *P. juliflora* plots indicates enhanced evaporation losses, though differences in soil water content may also be due to varying rate of transpiration among the species. Thus, spreading canopy of *Prosopis juliflora* and *Calligonum polygonoides* covered soil the best. The combination of *C. polygonoides* with *C. angustifolia* was best to control sand drift which resulted from active growth phase of *C. angustifolia* during peak supper period when maximum sand drift took place. At the age of 50 months, *A. tortilis* produced 5.2 tonnes ha^{-1} fuel wood as compared to 7.00 tones ha^{-1} from *P. juliflora* and 7.15 tones ha^{-1} from *Calligonum polygonoides*. *C. polygonoides* produced the highest biomass in the form of fuel wood utilizing minimum amount of soil water (Table 2). Usable woody species suitable for the region are *Calligonum polygonoides, Lycium barbarum, A. tortilis* and others such as *Calotropis* sp., *Balanites aegiptiaca*, *Prosopis* sps., *Casuarina equisetifolia*, *Ziziphus* sp, *Acacia* sp., etc. Introduction of under shrubs like *Cassia angustifolia* (medicinally important species) and locally available grass species (for fodder production) helps in early stabilization of dune and provides additional income from these.

Table 2. Biomass of the 50 months old tree seedlings (kg seedling^{-1}) on a shifting dune.

Tree species	V type	Above biomass	Root biomass	Total biomass	Surface biomass (tones ha^{-1} fresh weight)		
		D wt	D wt	D wt	1998	1999	2000
A. tortilis	*CA*	13.3	4.1	17.4	3.97	3.80	3.70
	CC	13.0	4.0	17.0	0.97	1.40	1.29
	CL	13.2	3.8	17.0	-	0.30	0.46
P. juliflora	CA	18.1	8.0	26.1	5.17	4.98	4.06
	CC	16.6	6.6	23.2	1.42	1.77	1.56
	CL	17.8	6.6	24.4	-	0.35	0.49

C. polygonoides	CA	18.0	9.8	27.8	7.28	7.35	6.46
	CC	18.7	8.9	23.6	1.93	2.57	2.19
	CL	16.9	7.8	24.7	-	0.58	0.60

V type: vegetation type; CA: *Cassia angustifolia*; CC: *Cenchurus ciliaris*; CL: Control; D wt: Dry weight.

Cenchrus ciliaris produced green fodder of 1.22 tonnes ha^{-1} $year^{-1}$ with *A. tortilis*, 1.58 tones ha^{-1} $year^{-1}$ with *P. juliflora* and 2.23 tones ha^{-1} $year^{-1}$ with *C. polygonoides*. *Cassia angustifolia* produced dry leaves of 0.76 tones ha^{-1} $year^{-1}$ with *A. tortilis*, 0.96 tones ha^{-1} $year^{-1}$ with *P. juliflora* and 1.39 tones ha^{-1} $year^{-1}$ with *C. polygonoides* (Table 3). Considering the economic benefits from *C. angustifolia* leaves, a farmer can get Rs 16720 ha^{-1} $year^{-1}$ from *C. polygonoides* plot as compared to 9120 ha^{-1} $year^{-1}$ from *A. tortilis* plot (Singh, *et al.*, 2003).

Table 3. Biomass (tones ha^{-1}) of *Cassia angustifolia* under different tree species.

Tree species	Nov 1998		May 1999		Nov 2000		Mean		Cost (Rs ha^{-1})
	FL	DL	FL	DL	FL	DL	FL	DL	@ Rs 12 kg^{-1}
A. tortilis	1.47	0.76	1.43	0.80	1.25	0.71	1.38	0.76	9120
P. juliflora	1.88	0.97	1.85	1.04	1.53	0.87	1.75	0.96	11520
C. polygonoides	2.68	1.39	2.80	1.56	2.15	1.23	2.54	1.39	16720

FL: Fresh weight leaves; DL: Dry weight leaves

Carbon benefits on tree integration

The study in agroforestry indicated the usefulness of tree species in improving the carbon stock in the Indian desert. *E. officinalis* was most effective in enhancing C sequestration as compared to *H. binata* and *C. mopane* (Singh, 2005). Higher litter addition had positive influence on SOC accumulation. Adverse climatic conditions negatively affect carbon sequestration in arid region, which could be improved through integration of tree species in agricultural system to increase overall production along with benefits of carbon credits.

In the sand dune stabilization, carbon sequestration was 3.72 tones ha^{-1} with *A. tortilis*, 5.24 tones ha^{-1} with *P. juliflora* and 5.66 tones ha^{-1} with *Calligonum polygonoides*. There was an increase in SOM due to plantation and surface vegetation. *C. polygonoides* was the best neighbour with better soil water status underneath and highest facilitating influences than *P. juliflora* and *A. tortilis*. *Calligonum polygonoides* was best neighbour increasing the *C. angustifolia* population and therefore, effectively stabilized the dunes and increased biodiversity and improved soil carbon status.

Conclusion and recommendations

Tree integration has multi-facet benefits involving product benefits along with aesthetic benefits and environmental improvement. It increases overall productivity of per unit area. Though there may be crop reduction resulting from tree integration but that can be compensated by benefits provided by trees. However, for effective mobilization of mass community and their involvement in afforestation or tree integration in agriculture land following points may be considered:

- Effective communication through media about economic benefits like production of wood, fuel, fodder, and fruit can improve the livelihood of the rural on tree integration in agriculture land.
- Effective training and exposure/ exchange visits to other sites for mass awareness.
- Involving people in tree growing through people participation and benefits sharing.

- Promoting income generation activities and livelihoods along with tree plantation.
- Providing incentives to the farmers on crop loss resulting from tree integration.
- Sharing of benefits on carbon sequestration/ increasing carbon stock (standing green living carbon in the form of carbon credits).

References

Blevines, R.L. and Frye, W.W. 1993. Conservation tillage: An ecological approach in soil management. Advances in Agronomy, 51: 33-78.

Bockheim, J.G.; Jespen, E.A. and Heisey, D.M. 1991. Nutrient dynamics in decomposing leaf litter of four tree species on a sandy soil in north western Wisconsin. *Can. J. For. Res.*, 21 803-812.

Brown, S.; Hall, C.A.S.; Knabe, W., Raich, J.; Trexle, M.C. and Woomer P. 1993. Tropical forests: their past, present and potential future role in the terrestrial carbon budget. *Water, Air and Soil Pollut.*, 70: 71-94.

Dixon, R.K. 1995. Agroforestry systems: sources or sinks of greenhouse gases? *Agroforestry Systems*, 31: 99-116.

Gupta, G.N.; Singh, G. and Kachwaha, G.R. 1998. Performance of *Prosopis cineraria* and associated crops under varying spacing regimes. *Agroforestry Systems*, 40: 149-157.

Havelin, J.L.; Kissel, D.E., Maddux, L.D.; Classe, M.M. and Long J.H. 1990. Crop rotation and tillage effects on soil organic carbon and nitrogen. *Soil Sci. Soc. Am. J.*, 54: 448-452.

Lal, R. 1989. Conservation tillage for sustainable agriculture tropics vs. temperate environment. *Advances in Agronomy* 42: 186-197.

Meentemeyer, V. and Berg, B. 1986. Regional variation in rate of mass loss of *Pinus sylvestris* needle litter in Swedish pine forest as influenced by climate and litter quality. *Scand. J For. Res.*, 1: 167-180.

Sanchez, P.A. and Benites J.R. 1987. Low-input cropping for acid soils of the humid tropics. *Science*, 238: 1521-1527.

Sharma, R.; Sharma, E. and Purohit, R. 1997. Cardamom, Mandrain and nitrogen fixing trees in agroforestry systems in India's Himalayan region. I. Soil nutrient dynamics. *Agroforestry Systems*, 35: 255-268.

Singh, G. 2005. Carbon sequestration under an agri-silvicultural system in the arid region. *Indian Forester* 131(4): 543-552.

Singh, G. and Rathod T.R. 2002. Plant growth, biomass production and soil water dynamics in a shifting dune of Indian desert. *Forest Ecology and Management*, 171: 309-320.

Singh, G.; Bala N.; Kuppusamy, V. and Rathod, T.R. 2003. Adaptability and productivity of *Cassia angustifolia* in sandy soil of Indian Desert. *Indian Forester*, 129(2): 213-223.

Singh, G.; Bala N.; Mutha S.; Rathod T.R. and Limba N.K. 2004. Biomass production of *Tecomella undulata* agroforestry in arid India. *Biological Agriculture & Horticulture*, 22(2): 205-216.

Singh, G.; Mutha S.; Bala N.; Rathod T.R.; Bohra N.K. and Kuchhawaha G.R. 2005. Growth and productivity of *Tecomella undulata* based on an agroforestry system in Indian desert. Forests, Trees and Livelihood, 15(1): 89-102.

Stump, L.M. and Binkley D. 1993. Relationship between litter quality and nitrogen availability in Rocky Mountain forests. *Can. J. For. Res.*, 23: 492-502.

FORESTS OF DADRA AND NAGAR HAVELI

S.S. Kandpal
Deputy Conservator of Forests, Daman and Diu.,
Fort Area Moti Daman, Daman.

ABSTRACT

Dadra & Nagar Haveli was liberated on 2nd August 1954 from the Portuguese regime. After its liberation, the administration was run by a local body known as free Dadra & Nagar Haveli administration till its merger with Indian Union on 11th August 1961. This small territory with an area of 491 sq. kms is having more than 75% tribal population with their vibrant art and culture.

Silvassa, the capital of Dadra & Nagar Haveli (DNH) derives its name from the word "SILVA" meaning wood. Out of total of 19,967 ha. (i.e. 40% of total geographical area) of Forest land, 9000 ha is under Wildlife Sanctuary and remaining is Reserve Forest. The forest area is spread over 58 villages out of 70 villages of Dadra & Nagar Haveli. Tribal population of DNH depends on forests for meeting their day to day needs. The Forest Type is South Indian Tropical Moist Deciduous (3BC2).

There are two projects of importance. The Lion Safari, Vasona and Satmalia Deer Park. The Lion Safari was started in 2002, with three Asiatic Lions brought from Junagarh Zoo. The main objective of the Lion Safari is to complement and strengthen the national effort in conservation of the rich biodiversity of the country, particularly the wildlife fauna.

Satmalia Deer Park extends over an area of 300 ha. of chain-link fenced natural forest. Species like Sambar, Nilgai, Spotted deer, Four horned antelopes are released in the area. The area is currently having a population of 100 animals.

History and Culture of Dadra & Nagar Haveli

Dadra & Nagar Haveli was liberated on 2nd August 1954 from the Portuguese regime. After its liberation, the administration was run by a local body known as free Dadra & Nagar Haveli administration till its merger with Indian Union on 11th August 1961. This small territory with an area of 491 sq. kms is having more than 75% tribal population with their vibrant art and culture. Dhodia (including Halpati), Kathodi, Kokna, Koli Dhor (including Kolgha), Naika and Varli are the tribes of the territory. Out of these, main groups are Dhodia, Kokna & Varli.

The terrain is part hilly with Sahyadri Ranges and rest is plane. Damanganga River flows into the territory with Piparia, Kolak and Sakartod as its tributaries. The territory is having 70 villages and Silvassa-Amli township.

Accessibility

Nearest Rail head is Vapi at a distance of 18 Kms on Mumbai-Delhi route. Mumbai-Delhi National Highway No.8 connects Dadra & Nagar Haveli via Vapi.

General Statistics

Area	:	491 Sq. Kms.
Mean Annual Rainfall	:	2000 to 2500 mm.
Climate	:	Pleasant; Warm in summer, Cool during winters
Temperature	:	39°C Maximum and 11°C Minimum
Population	:	2,20,451 (2001 Census).
Language Spoken	:	Gujarati, Marathi, English and Hindi.
Decade Growth Rate	:	59.20% (1991-2001). (*against 21.34 at*

		country level).
Density per Sq. Km	:	449 Persons (2001). (*against 324 persons at country level*).
Sex Ratio	:	811 (Females per 1000 Males). (*against 923 at country level*).
Birth Rate	:	33.21 (Per '000). (*against 25.4 at country level*).
Death Rate	:	4.95 (Per '000). (*against 8.4 at country level*).
Infant mortality rate		
Per '000 live births	:	11.46 (*against 66 at country level*).
Maternal Death Rate	:	1.13. (Per '000)
Literacy Rate	:	60.03% (*against 65% at country level*).
Male	:	73.32% (*against 76% at country level*).
Female	:	42.99% (*against 54 % at country level*).
Urban /	:	50546 (Urban)
Rural Population	:	169995 (Rural)
Distribution		

Forests and Wildlife

Forests

Silvassa, the capital of Dadra & Nagar Haveli (DNH) derives its name from the word "SILVA" meaning wood. Out of total of 19,967 ha. (*i.e.* 40% of total geographical area) of Forest land, 9000 ha is under Wildlife Sanctuary and remaining is Reserve Forest. The forest area is spread over 58 villages out of 70 villages of Dadra & Nagar Haveli. Tribal population of DNH depends on forests for meeting their day to day needs.

The forest type is South Indian Tropical Moist Deciduous (3BC2). The top canopy comprises Teak (*Tectona grandis*), Sadad (*Terminalia tomentosa*), Haldwan (*Adina cordifolia*), Tiwas (*Ougeinia oojeinensis*), Bondar (*Lagerstroemia parviflora*), Mahua (*Madhuca indica*), Khair (*Acacia *catechu*), Kanti (*Acacia ferruginea*), Shisham (*Dalbergia latifolia*), Jambul (*Syzygium cumini*).

The middle canopy comprises Kumbhio (*Careya arborea*), Timru (*Diospyros melanoxylon*), Kharoti (*Ficus asperrima*), Awla (*Emblica officinalis*), Ashitro (*Bauhinia racemosa*), Khakher (*Butea monosperma*), Umbar (*Ficus glomerata*), Manvel vas (*Dendrocalamus strictus*), Katas vas (*Bambusa arundinacea*).

The undergrowth is having Karvand (*Carissa carandas*), Nirgudi (*Vitex negundo*), Ukshi (*Calycopteris floribunda*), Rui (*Calotropis gigantea*) and ground cover comprises Randhendi (*Thespesia macrophylla*), Ranambadi (*Hibiscus furcatus*), Chikta (*Desmodium pulchelum*). Bokarvel (*Combretum ovalifolium*), Garvel (*Tinospora cordifolia*), Nandalvel (*Vitis repanda*) are main climbers.

Indian Forest Act, 1927 (Gujarat Amendment applicable), Wildlife Protection Act 1972, Goa Daman & Diu Preservation of Tree Act, Forest (Conservation) Act – 1980 together with all Central Government and Supreme Court guidelines are followed in the territory while dealing with the matters related to forests.

Revised Working Plan for Forests and Management Plan for Sanctuary areas are under preparation with the help of ICFRE, Dehradun.

Joint Forest Management

It was realized that forests of Dadra & Nagar Haveli are getting degraded due to excessive pressure, especially by grazing, removal of firewood and timber, seasonal cultivation etc. It was also felt that Forest Department of UT administration alone cannot ensure protection of forests without the co-operation of villagers/local people.

Therefore as per the guidelines issued by the Ministry of Environment & Forests, the UT Administration has issued notification for initiation of Joint Forest Management. The same has been initiated in five villages and it is intended to spread it to other villages from the experience of these villages.

Wildlife

The main Wildlife found in the territory are Panther (*Panthera pardus*), Jungle Cat (*Felis chaus*), Mongoose (*Herpstes edwardsi*), Fox (*Vulpes bengalensis*), Hyaena (*Hyaena hyaena*), Chousingha (*Tetracerus quadricarnis*), Sambhar (*Cervus unicolar*), Nilgai (*Boselaphus tragocamelus*), Chital (*Axis axis*), Hare (*Lepus nigricollis*), Bonnet Macaque (*Macaca radiate*), Peacock (*Pavo cristatus*), Red vented bulbul (*Pycnonotus cafer*), Common Cobra (*Naja tripudians*) and Python (*Python molrus*).

Estimation of Wildlife (2002-03)

Name of Animal	Wild	Captive
Lion	--	02
Panther	08	--
Hyaena	06	01
Chausinga	09	04
Crocodilus goradim	02	--
Civet cat	02	--
Wild boar	01	--
Jackal	04	--
Mangoose	--	04
Hare	03	08
Nilgai	14	--
Cheetal	16	05
Sambhar	31	04

Rights and Concessions

No concessions were available to local people during Portuguese regime on forests or over its produce. The present rights involve right of cutting twiges, branches, collection of fuel wood (one head load). Locals are also allowed to graze their animals in Reserve Forest areas except in regeneration areas. The local people also collect honey, mahuwa etc. from forest areas. Following grants used to be given till 1996 to local people for house hold uses in relation to timber.

Free grant of timber for construction of house ⟶ 2.5 cm.
Free grant of timber for repair of house ⟶ 0.25 cm.
Free grant of timber for agricultural implements ⟶ 0.05 cm.
Concessional grant of timber at an interval of 10 yrs. ⟶ 2.5 cm.

As there is no timber operation nowadays, therefore these rights are not in operation/executed.

Forestry Budget

A typical statement of budget allocation

Tenth Five Year Plan Outlay to the UT- **30400 Lakhs**
Allocation to Forests - **1200 Lakhs**

Further following assistance is received under Centrally Sponsored Scheme from MoEF for Sanctuaries:

2002-03 **15.00 Lakhs**
2003-04 **21.00 Lakhs**

Approximately Rs. 162.00 Lakhs is allocated under non-plan every year.

Afforestation Targets and Achievements

During ninth Five Year Plan (1997-2002), following are the targets set and achieved:

Name of the Scheme	Target	Achievement
Forest Conservation Regeneration and Development.	300 ha.	567 ha.
Social & Farm Forestry	1025 ha.	1845 ha.

Distribution of Seedlings	25 lakhs	28.5 lakhs
Compensatory afforestation	--	197.15 ha.
Total area	1325 ha.	2607 ha

Special Projects of Importance

Lion Safari, Vasona

Lion Safari was started in 2002, with three Asiatic Lions brought from Junagarh Zoo. The main objective of the Lion Safari is to complement and strengthen the national effort in conservation of the rich biodiversity of the country, particularly the wildlife fauna. The following are the main aims and objectives:

a. The conservation of endangered species of asiatic lion for breeding in existing conditions and raise stock for their rehabilitation in the wild.

b. To inspire among safari visitors empathy for wildlife, an understanding and awareness about the need for conservation of natural resources and for maintaining the ecological balance.

c. To provide opportunity for scientific studies useful for conservation in general and creation of data base for in-situ and ex-situ conservation studies.

The lion safari is spread over an area of 20 ha. at Vasona. The entire area is protected with seven meter high chain link wall having 2700 meter length. The internal road network is 2.6 Kms long and touches all-important areas of the Lion Safari.

The lion safari was having three lions (one male & two females) till last month. One female died last month due to illness. The proposal to get two more lions is under active consideration. This lion safari is a big tourist attraction. For the last two years, nearly 50,000 people visit lion safari every year.

Satmalia Deer Park

Satmalia Deer Park extends over an area of 300 ha. of chain-link fenced natural forest. Species like Sambar, Nilgai, Spotted deer, Four horned antelopes are released in that area. Nearly ten years back, due to dwindling population of the wild animals in the territory, a project of captive breeding of these animals was started and the animals born thus were released in the wild in places such as Satmaliya. The present population of spotted deer, nilgai, sambhar and four horned antelopes in the enclosure is nearly 100.

A large number of tourist and trackers visit the area both for love to nature, fun and adventure. Further Forest Department is tying up with resorts at Silvassa, to take their package tourists to lion safari and Satmalia Deer Park.

ENCROACHMENTS

TERAM CULTIVATION

History of "Teram" dates back to Portuguese regime when forest lands were given out to people for cultivation on lease annually. The lease was renewed every year. In 1969, the forests of Dadra and Nagar Haveli were declared Reserved Forests but "Teram Plots" were not recognized as they honey comb the forests. Forest Conservation Act 1980 put an end to the practice of lease, however practice of cultivation continued to exist on ground. Several literature consulted were made to Government of India since 1980 for settling the matter once and for all but the matter still remains to be sorted out.

Basic Statistics of Teram Encroachment

There are a total of 845 plots held by 792 cultivators covering an area of 613.30 ha. Out of 845 plots, 299 plots are within the wildlife sanctuary and 546 are within the reserved forest area. These cultivators are also having land holding (Malkiland) in addition to this cultivation. Out of 792 cultivators 140 are land less, 142 have land upto 1 ha., 192 have land between 1 – 2 ha., 106 have land between 2 – 3 ha., 61 have land between 3 – 4 ha., 19 have land between 4 – 5 ha., and 51 have land more than 5 ha.

Following are two broad alternatives of this vexes issue

Alternative – I: In-situ settlement of cultivators and regularization of encroached land. This may satisfy all but requires the permission of Govt of India under Forest Conservation Act, 1980 and also requires compensatory afforestation etc. to be complied with. However, this will create small pox like scars on forests. Providing network of roads, for access and other basic amenities would be difficult and will damage the forests further.

Alternative – II: All the cultivators should be shifted out of the forests and settled in one or more compact blocks. This will free the forests from encroachment but U.T. Administration is not having such large area available, and Teram encroachers also do not want to shift.

The important decision/committee recommendations about Teram cultivators are as follows:

A. On 07/04/1984 in a meeting taken by the then Home Secretary, which was attended by the Secretary (Agriculture), IG Forests and Joint Secretaries from Ministry of Home and Agriculture besides DCF Dadra and Nagar Haveli, the following recommendations were made

   1. Teram cultivators be allowed to continue the lease under the existing arrangements on annual renewal basis provisionally for a period of 10 years, provided they take up tree cultivation in a phased manner in those areas.

   2. Teram cultivators should be given sufficient incentives, like the social security through plantations under social forestry programmes so as to make tree planting attractive in Teram plots.

Usufructs of the plantations should be allowed to the cultivators.

3. Local staff should motivate Teram cultivators suitably and also provide free seedlings and technical guidance.
4. In the meantime, suitable alternative areas should be identified where cultivators should be re-settled in a phased manner. Lease should be inheritable but not to be transferred or sold.
5. The lease should be terminated, if the cultivators do not take up tree plantations in the plots within the specified period.

B. The Inter-Ministerial Study Team consisting of Shri K. M. Chadha, Joint Secretary, Ministry of Environment & Forests, Dy Secretary, Ministry of Home Affairs and Conservator of Forests (Regional Office), Bhopal visited Silvassa on 20.01.1988, and based on the extensive discussions and field visits, made the following observations:

   a. There are about 800 small Teram plots under agricultural crop inside forest area. It was thought advisable to consolidate these small holdings outside or on the fringes of forest area to make them about 8 to 10 large units and all the agricultural plots from within the forest area should be removed.
   b. The new plots to be given should be equivalent in area to the plots being surrendered, but not including the encroachments by the Teram cultivators.
   c. A detailed study will have to be done by Dy. Commissioner, Silvassa and DCF, Silvassa to earmark areas where the Teram cultivators could be reallocated agricultural land.

C. In 1990s Dadra and Nagar Haveli administration took the following stand:

i. If the Teram cultivators are settled in-situ, equivalent non-forestland will have to be transferred to the Forest Department and the availability of such lands outside the forest area is in question. Even if the land is to be acquired by payment of compensation, the cost thereof and the cost of compensatory afforestation as required under FC Act will be enormous.

ii. Even those who are landless or have malkiland less than 1 ha., the land required will be 213 ha., the cost of such land will be prohibitive.

iii. De-reservation of Teram plot is neither desirable nor feasible, as it will make forest working difficult and result in further encroachments and illegalities.

It was recommended that plantations should be raised on the plots by giving free inputs on the promise that usufructs be given to the landless Teram cultivators. This would serve the purpose of forestry as well as the Teram plot holders.

D. An affidavit was filed in 1997 in Supreme Court of India informing the Court of the status of Teram cultivation in the territory.

E. Another affidavit was filed in Supreme Court in year 2002 recommending that till the UT Administration finds a viable solution to this problem, the Teram cultivators should be allowed to continue their cultivation.

CHORTI CULTIVATION

Due to non-availability of sufficient land for agriculture the tribals in the interior forests some times also resort to unauthorized cultivation in forest areas. The forest areas being rich in nutrients are chosen mostly for raising paddy seedlings for duration of 3 to 6 months. These encroachments are seasonal in nature and the encroachers do not necessarily cultivate the same area next year. To tackle these encroachments offence cases are booked under IFA.

Following is the unauthorized cultivation recorded during a span of five years:

Year		Area under encroachment (Ha.)
1999	-	83.74
2000	-	34.10
2001	-	5.386
2002	-	39.8
2003	-	16.55

For checking encroachments Forest Department is fencing the sensitive forest areas and Joint Forest Management is also proposed to be implemented to elicit involvement of people for tackling the issue of encroachment. Further the administration is of the considered view not to allow such type of encroachments.

Conclusion

Forest Department, Dadra & Nagar Haveli is committed to protect the forests by way of creating awareness among people.

Centres like Lion Safari Vasona and Deer Park Satmalia help to arouse a feeling of love for Forest & Wildlife. Infact, these two centres will go a long way for creating a people and forests interface.

While dealing with old encroachments, a view of toleration and persuasion from the side of Forest Department has been appreciated by local public and which also has ultimately helped to serve the cause of forestry in the territory.

APPLICATIONS OF REMOTE SENSING AND GEOGRAPHIC INFORMATION SYSTEM TECHNIQUES IN FOREST STUDIES

Rakesh Paliwal and J. R. Sharma
RRSSC / ISRO / Dept. of Space, CAZRI Campus -Jodhpur

ABSTRACT

Sound management of forest resources from local level to national level requires good assessment of situation in different seasons over a period of time. In order to produce good assessment, it is necessary to have timely and reliable data and information. Remote sensing is a science of obtaining information about an object or phenomenon, from a distance, without actually coming into physical contact with the target, one such technology, widely used for studying different aspects of forest resources. Satellite remote sensing provides multi-spectral, multi-spatial, multi-temporal and multi-sensor data, which can be used in forest resources inventory, mapping, monitoring and their management.

There are various applications of remote sensing *viz.*; Generation of Forest Resource Inventory, Forest Cover and Density Classification, Development of Habitat Diversity Indicators for Landscape Assessment, Change Detection, Forest Productivity/Volume Estimation, Forest Fire Detection, Tree Height using LIDAR Data etc.

The present paper demonstrates through three case studies Hunsur District, Karnataka; Delhi Ridge Forest and Forest Crown Density, Jhalawar, Rajasthan; some of the applications *viz.*; forest volume estimation, Analysis of landscape based habitat diversity indices and Forest Change Detection.

Introduction

Forest is one of the major sources of carbon flux. Forest plays an important role in regulating the environment and its processes. Hence, forest is an important component of Biosphere. In the social aspect, forest is also important to mankind as it provides the basic necessities like food, fuel and shelter. With increasing population of mankind, the equilibrium between supply and demand of forest produce is getting disturbed. However, to avail the goods and services of forest ecosystem in sustainable manner, the forest management systems are evolved. The challenging task for the forest management system is monitoring and inventorization of forests.

The different objects on the surface of the earth reflect different amounts of energy in different wavelength of the Electromagnetic Spectrum. Detection and measurement of these spectral responses and spectral signature is called the process of remote sensing. These spectral responses and spectral signature enable us to identify and classify the surface objects through visual interpretation techniques and/or digital analysis using computers and supported by ground verifications / measurements and there by improving the accuracy of thematic content of the analysis.

Further, remotely sensed data is compatible and can be integrated with most of the other data generated from conventional and ground measurement systems using Geographic Information System (GIS) for finding optimal solutions for effective management of forest resources and environment for sustainable development. These tools used in combination bring flexibility into the data analysis. Using GIS, data from different sources and scales can be integrated within a geo-referenced environment providing changes at pixel level over a period of time.

Generally, in remote sensing based forest studies, the forest cover is divided into evergreen / semi evergreen, deciduous (dry / moist / thorn), scrub forest, forest blank and forest plantations based on the forest types available. Further, it is also divided as per density into dense/closed (canopy density >40%), open (canopy density 10-40%), scrub forest (canopy density <10%) and forest blank (almost no vegetation) based on the canopy density.

The remote sensing is a useful tool for achieving the task described above. The synoptic, temporal, spatial and radiometric feature of remote sensing provides an effective tool to estimate the various characteristic of forests. The temporal aspect of remote sensing can provide the regular interval monitoring and inventorization of forests. Remote sensing can also provide quantitative information on bio-physical variables such as vegetation structure, growing stock, productivity and transpiration (Franklin and Strahler, 1988).

In the present study, first case study shows the possibility of forest volume estimation for the part of Hunsur Distirct of Karnataka using the modeling approach of the remotely sensed data, collateral data etc, which is based on regression technique. The second case study investigates the relationship between landscape pattern and ecological processes of Ridge forest of Delhi. This process lead to develop spatial landscape indices using remote sensing data which would quantify the landscape pattern to reflect the ability of the landscape to support certain ecosystem functions. The third case study demonstrates the change detection analysis for Jhalawar District of Rajasthan.

Case Study-I

Forest Volume Estimation for Part of Hunsur Forest

The various modeling techniques for determination of tree structure is used which include Regression Model (Least-square) (Ardo, 1992; Jaques & Cihlar, 1980 and Oza *et al.*, 1996), Geometric-Optical Canopy Reflectance Model, Multivariate Adaptive Regression Model (MARS), Generalized Linear Models (GLMs) (Guisan, *et al.*, 2000), Artificial Neural Networks (ANNs) etc. In the present study, the multiple regression model is used to estimate tree structure variables (Paliwal, 2003).

Tree Structure Attributes

Measurement of tree in general, are made for following variables which are essential attributes for Biomass/Yield estimation:

- Tree age
- Tree Height (H)
- Diameter at breast height (DBH)
- Number of trees
- Canopy Structure (mainly canopy diameter and crown area)

Study Area

The study area comprises part of Hunsur Forest Division, Karnataka, India. It is located on the southeastern slope of the Western Ghats. It is part of Nagarhole Wild Life Sanctuary (WLS). It is located between 11°52" & 12°21" N latitudes and between 75°47" & 76°22" E longitudes. The location map of Hunsur District is shown in Figure 1. The altitude varies roughly between 790 and 1220 m above *msl*. The average annual rainfall in the area varies between 65 & 180 cm. Degraded forest and scrubs were found in the east and moist deciduous forest in the west. The climate is generally tropical. The shade temperature varies between 55°F and 95°F. The underlying rock is of granitic gneiss, mixed here and there with charneekite or Nilgiri gneiss. In the natural tropical dry deciduous forest, *Tectona grandis, Terminalia tomentosa, Pterocarpus marsupium* etc. occur as dominant species.

Figure -1 Location Map

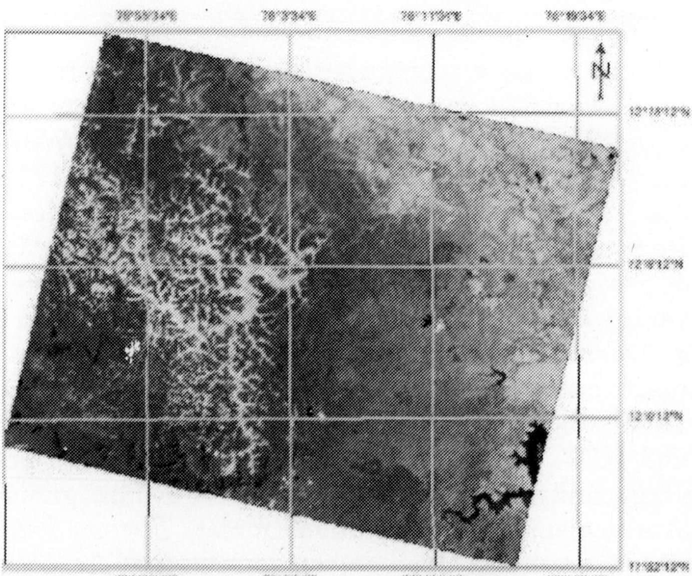

Figure - 2 Satellite Image (FCC)

Field Data

In this study, twenty-six samples are analyzed (n=26) of 30 x 30 m size. The samples belonged to evergreen, semi evergreen, mixed moist deciduous, dry deciduous and degraded forests types. These forest classes can be seen in the satellite data (False Color Composite) shown in Figure-2. All trees present in each sample were enumerated for basal circumference (BC), diameter at breast height (DBH), total tree height (H) and canopy diameter (CD). The details of tree measurements (mean values of DBH, CD, H, and BC) and forests types for sample locations are given in Table 1.

Table 1. Tree Structure Measurements

Sample No.	Trees	DBH	CD	H	BC	Forests type
1	10	40.06	7.93	22.65	81.43	Dry deciduous
2	12	24.08	5.82	20.31	103.88	Dry deciduous
3	10	28.11	7.35	18.17	94.25	Degraded Forest
4	14	20.39	4.86	10.21	141.5	Dry deciduous
5	11	47.93	8.23	24.36	99.89	Dry deciduous
6	18	24.83	5.98	13.88	125.23	Dry deciduous
7	11	35.28	7.03	22.95	159.11	Dry deciduous
8	17	22.54	4.23	15.57	122.89	Dry deciduous
9	12	35.16	7.98	21.69	170	Dry deciduous
10	9	32.04	7.07	22.56	131.36	Dry deciduous
11	9	41.54	8.16	19.61	78.65	Dry deciduous
12	11	45.39	9.83	20.63	132.92	Dry deciduous
13	9	53.12	9.22	26.06	169.11	Dry deciduous
14	12	42.93	8.75	23.96	168.17	Dry deciduous
15	13	41.88	9.57	25.04	198	Mixed Moist Deciduous
16	14	41.68	8.87	21.34	104.21	Mixed Moist Deciduous

17	17	25.32	6.01	13.8	108.92	Dry deciduous
18	18	20.49	5.23	14.42	82.28	Mixed Moist Deciduous
19	13	31.89	6.74	15.77	162.46	Mixed Moist Deciduous
20	8	27.59	6.41	10.97	159.29	Mixed Moist Deciduous
21	12	28.98	6.55	19.58	97.88	Mixed Moist Deciduous
22	9	29.57	8.36	14.31	106.33	Semi Ever Green
23	10	44.76	9.21	21.72	118.67	Semi Ever Green
24	4	76	13.4	30.2	195.63	Semi Ever Green
25	7	51.1	9.06	18.57	277	Mixed Moist Deciduous
26	9	71.97	13.69	24.75	194.29	Semi Ever Green

Satellite Data

In the present study, three satellite data sets are used *viz.*, Landsat-4 Thematic Mapper and two sets of IRS-1B LISS-II. The details of these are presented in Table 2.

Table 2. Details of Satellite Data

S. N.	Satellite Sensor	Spectral Bands (μm)	Date	Path/Row
1.	Landsat-4 Thematic Mapper (TM)	B1: 0.45-0.52 B2: 0.52-0.60 B3: 0.63-0.69 B4: 0.76-0.90	12th March, 1987	145/50
2.	IRS-1B LISS-II	B1: 0.45-0.52 B2: 0.52-0.59 B3: 0.62-0.68 B4: 0.77-0.86	30th January, 1997 24th March, 1998	26/60

Methodology

In literature, many studies on the subject are found based on regression models. Based on its limitations and advancements in modelling techniques, many improved models have been evolved. In the present study, the conventional multiple regression model is used. To design regression models, the model variables are defined, which include the tree structure variables, topographical variables and remote sensing data (reflectance in various bands of sensor) as variables.

Forest Classification: Champion and Seth (1968) have classified the following forests in this region:

- West Coast Tropical Evergreen Forest (1A/c4)
- West Coast Tropical Semi-evergreen Forest (2A/c2)
- South Indian Tropical Moist Deciduous Forests (3A/c1)
- Southern Dry Deciduous Forests (4A/c2)
- Scrub Forests (5/5B/Ds1)

To identify the spatial extents of the above mentioned forest classes, the supervise classification technique with maximum likelihood classifier is used. The Landsat-4 TM satellite image is used with different classes of training sets. The classification is performed using ERDAS Imaging software. The broad forest cover and non-forest cover is shown in Figure 3. The classes identified are evergreen forest, semi-evergreen forest, moist deciduous forest, dry deciduous forest, dry shoal scrub, degraded forest, scrub forest, sand/gullied ravines and water body. The area estimates are derived for different classes and shown in Table 3. The overall accuracy of 95.67% is achieved.

Modelling

Correlation Analysis and Multiple Regression Model

To perform the multiple-regression, the response and predictor variables are decided. The main aim of the study is to estimate the biomass, which depends on DBH & H or CD or Basal area of the tree/crop. In the case of indirect approach of biomass estimation, preferred response variable has to be selected. In order to search the same, the magnitude of the relationship between spectral response of satellite data and tree structure variables need to be understood. For that the correlation matrices are generated using the STATGRAPHICS 2.6 software. After correlation analysis, the regression modeling is performed. The Table 4 shows the results of regression model.

Table 3. Area estimates of different classes in study area

Classes	Area (ha)	Area (%)
Tropical evergreen forest	25407.54	11.31
Tropical semi evergreen forest	36618.21	16.29
Moist Mixed Deciduous forest	20800.89	9.26
Tropical Dry Deciduous forest	52613.46	23.41
Dry Shola Scrub	7887.78	3.51
Degraded forest	7155.81	3.18
Non-Forest Lands	56101.5	24.96
Sand/ Gullied Ravines	15944.04	7.09
Water Body	2057.13	0.92
Cloud (Unclassified)	148.5	0.07
Total	224734.9	100.00

Table 4. Fitted regression models with CD-response variable

IRS-1B LISS-II January, 1997		
S.N.	Predictors/Independent Variables	R-SQ
1.	RATIO.b21	0.285
2.	RATIO.b13	0.252
3.	BAND.B1JanL2	0.270
4.	RATIO.b21, Slope, Aspect	0.359
IRS-1B LISS-II March, 1998		
1.	RATIO.b21	0.344

2.	RATIO.b14	0.295
3.	RATIO.b31	0.316
4.	RATIO.b41	0.276
5.	BAND.B1MarL2	0.356
6.	BAND.B1MarL2, Slope, Aspect	0.483
7.	RATIO.b21, Slope, Aspect	*0.492*
Lansat 4-TM March, 1987		
1.	RATIO.b43	0.359
2.	RATIO.b43, Slope, Aspect	0.465

Note: b21, b13, b41, b31, b13 and b43 represents the band-ratios e.g. b21=band2/band1

Note: Significant At 95 Percent Confidence Level

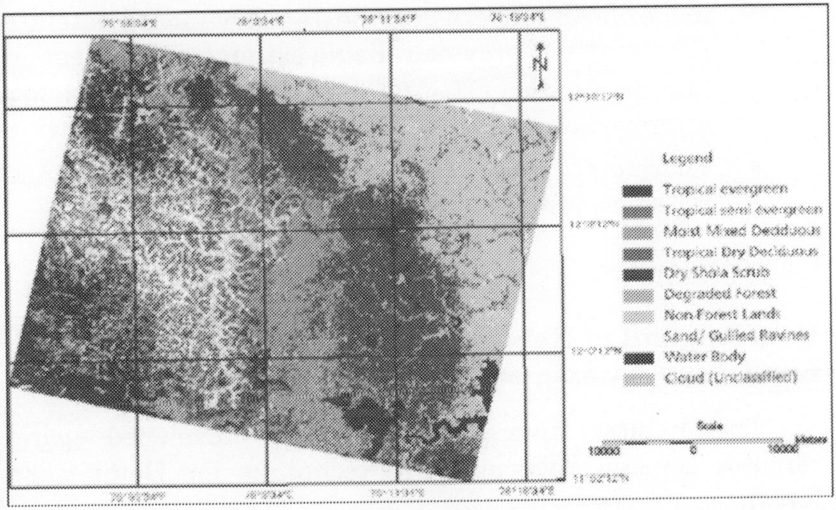

Figure 3. Classified Satellite Image

Conclusion

- In three spectral data sets, the maximum correlation value is observed for canopy diameter (CD) (0.52 - Jan 97, 0.60 - Mar 98, and 0.42 – TM 87). This infers that CD out of the other tree structure variable better relates the spectral information, which in turn could be used to estimate biomass.

- Basal circumference and DBH are the next variables to CD that contains the high correlation value with spectral response, but it cannot provide the volume/biomass estimates as the volume measures required the tree height information also. And it is observed that the relation between tree height and spectral response is very week.
- Hence, in the indirect approach of biomass/volume estimation, the canopy diameter plays an important role.
- In multiple regression model for January month, R^2 values (0.25, 0.27, 0.28 & 0.35) are observed as insignificant. Therefore, the January month satellite data may not provide reliable estimates. Whereas in March months, The R^2 values (0.48 & 0.49) are improved compared to January month.
- Therefore, these regression models can provide reliable estimates upto 48-49% for CD.

Case Study-II

Development of Habitat Diversity Indices for Landscape Assessment of Delhi Ridge Forest

The habitat diversity indices are developed using remotes sensing data and GIS technique for Delhi ridge forest.

The Delhi Ridge is the northern extension of Aravalli hill range. It is entering the National Capital Territory of Delhi in the Southeast at the Tughlakabad- Bhatti mines-Dera Mandi axis and terminating in the north near Wazirabad on the right bank of Yamuna river.

The Government of National Capital Territory of Delhi has declared four ridges as reserved forest.

1. Northern Ridge (Area of 87.0 hectares)
2. Central Ridge (Area of 864 hectares)
3. South - Central Ridge (Area of 626 hectares)
4. Southern Ridge (Area of 6200 hectares)

The Ridge and the vegetation are critical for the lifeline of the NCT-Delhi due to its following functions

- Assimilation of gaseous pollutants
- Groundwater recharge
- Prevention of erosion and siltation
- Maintenance of bio-diversity
- Aesthetic and recreational opportunities

Data

The Satellite data of IRS 1A LISS-II of September 1992 and IRS 1C LISS-III of October 1996 are used for development of indices. The Figure 4 shows the Satellite data (False Color Composite) of Ridge forest of Delhi for the year 1992 and 1996.

Methodology

Habitat Diversity Indices

To investigate relationships between landscape pattern and ecological processes, spatial landscape indices are developed which would quantify the landscape pattern to reflect the ability of the landscape to support certain ecosystem functions (Paliwal, 2001). To measure the landscape pattern, the following spatial indices are considered

- Simpson
- Diversity
- Evenness
- Dominance
- Fragmentation

- Fractal Dimension
- Edge length or edge density
- Contagion

The unsupervised classification is performed to know the saturation of classes and to develop diversity index for all the four ridges for two point of time. The other above mentioned indices are developed for both time period for all four ridge forest areas and the flow chart of the same is given in Figure 5.

Further these indices values are used for identifying the relationship among the habitat diversity indices using fuzzy if-then rule based model. The fuzzy number for each habitat index is derived and fuzzy rules are elicited through expert's opinion. The fuzzy rules are validated through iteration process. The validated model is then used for prediction. Table 5, Table 6, Table 7 and Table 8 show the observed and predicted values for combination of indices.

Based on the analysis of the habitat indices values using fuzzy-if then rule model, the ranking of the four ridges is as follows:

Ridge Forest	Rank
Northern Ridge	I
Central Ridge	III
South Central Ridge	II
Southern Ridge	IV

Table 5. Edge Density, Fractal Dimension & Diversity

Name of the Ridge	Observed value for Diversity	Predicted value for Diversity	Root Mean Square Error
1992			
Northern	2.9238	3.00	0.0762
Central	3.1318	3.00	0.1318
South-central	3.0559	3.00	0.0559
Southern	3.2006	3.25	0.0494

1996			
Northern	2.7155	2.875	0.1595
Central	3.0623	2.897	0.1653
South-central	3.1044	2.875	0.2294
Southern	3.2025	3.250	0.0475

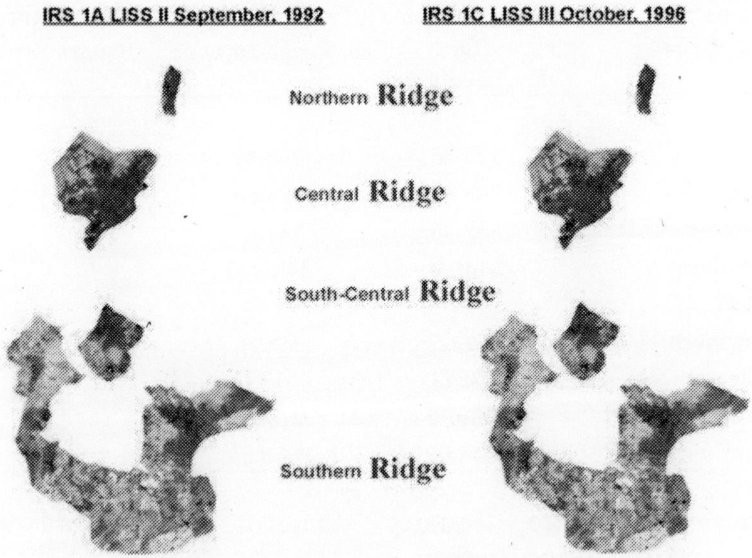

Figure 4. Satellite Date (FCC) of Ridge Forest of Delhi

Table 6. Edge Density, Fractal Dimension and Diversity (Simpson Index)

Name of the Ridge	Observed value for Diversity	Predicted value for Diversity (Simpson Index)	Root Mean Square Error
1992			
Northern	0.0629	0.0515	0.0114
Central	0.0461	0.0499	0.0038
South-central	0.0527	0.0511	0.0016
Southern	0.0414	0.0425	0.0011
1996			
Northern	0.0759	0.0637	0.0122
Central	0.0548	0.0615	0.0067

South-central	0.0521	0.0637	0.0116
Southern	0.0413	0.0425	0.0012

Table 7. Edge Density, Fractal Dimension and Fragmentation

Name of the Ridge	Observed value for Fragmentation	Predicted value for Fragmentation	Root Mean Square Error
1992			
Northern	103.43	100.00	3.43
Central	793.87	794.85	0.98
South-central	382.58	381.25	1.33
Southern	3446.9	3425.00	21.9
1996			
Northern	371.66	943.75	572.09
Central	2772.59	2515.44	257.15
South-central	1288.12	943.75	344.37
Southern	9844.69	9850.00	5.31

Table 8. Edge Density, Fractal Dimension and Contagion Index

Name of the Ridge	Observed value for Contagion Index	Predicted value for Contagion Index	Root Mean Square Error
1992			
Northern	50184.88	50250	65.12
Central	565893.63	557029.41	8864.22
South-central	238325.47	255375.00	17049.53
Southern	2951115.66	2950000.00	1115.66
1996			
Northern	144382.29	665625.00	521242.71
Central	1450799.5	964632.35	486167.15
South-central	597549.48	665625.00	68075.52
Southern	6108759.1	6125000.00	16240.9

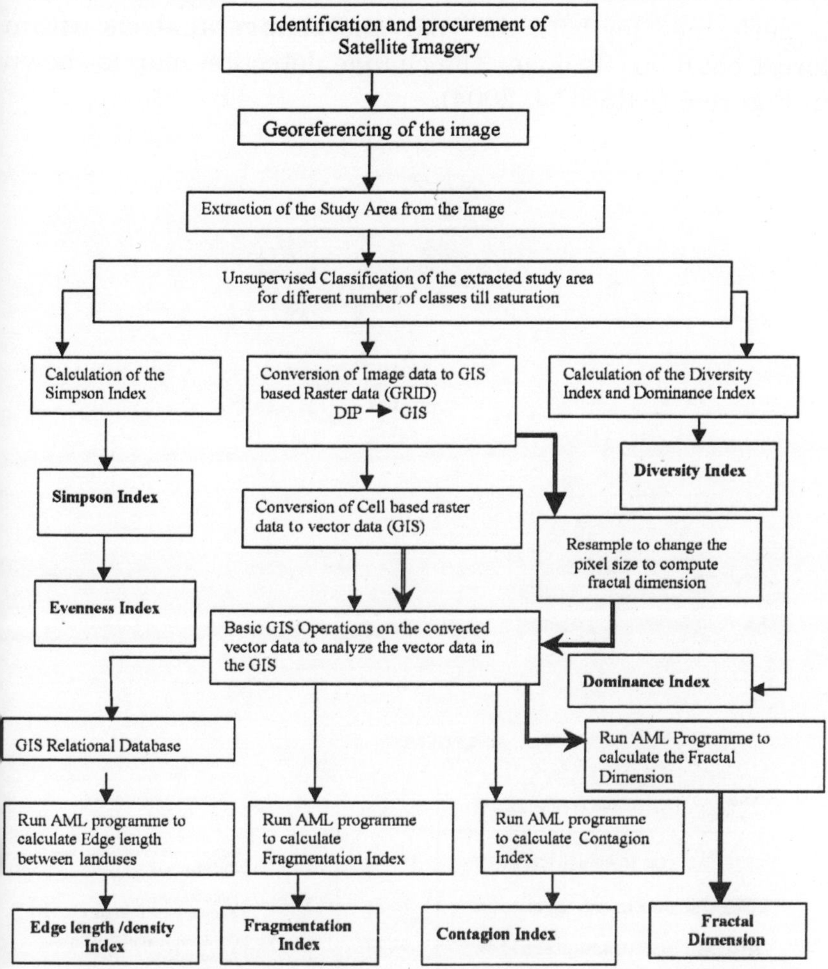

Figure 5. Method to calculate landscape pattern Indices

Case Study-III

Forest Crown Density Change Detection of Jhalawar District

In this case study, 1996-97 and 2000-01 satellite data (LANDSAT-TM and IRS 1C / 1D-LISS-III) are used for analyzing the change detection of the forest crown density. The digital image processing (maximum likelihood classifier) is performed for both the year and change analysis within forest boundary is done. The change detection map is shown in Figure-6 (RRSSC-J, 2004).

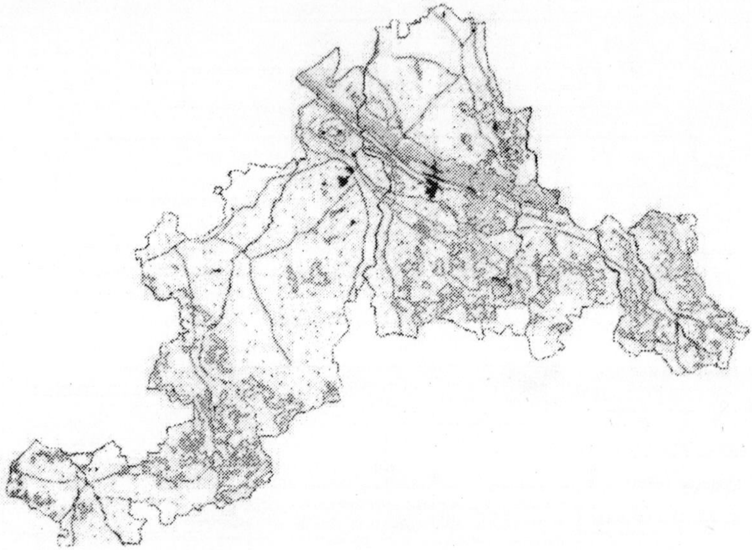

LEGEND

Map Symbol	Nature of Change	Area (Sq. km.)	% Change	Drivers / Reasons for Change
	FCD 10-40% to FCD 40-70%	143.50	70.26	Construction of Reservoir /plantations
	FCD 10-40% to FCD <10%	26.17	12.81	Less Rainfall and population pressure
	FCD 10-40% to FCD >70%	02.79	01.37	Construction of Reservoir /plantations
	FCD 40-70% to FCD 10-40%	00.11	00.05	Less Rainfall and population pressure
	FCD 40-70% to FCD >70%	07.83	03.84	Construction of Reservoir /plantations
	FCD <10% to FCD 10-40%	18.35	08.99	Construction of Reservoir /plantations
	FCD <10% to FCD 40-70%	05.29	02.59	Construction of Reservoir /plantations
	FCD >70% to FCD 40-70%	00.10	00.20	Less Rainfall and population pressure

Note : FCD - Forest Crown Density

Figure 6. Change Detection Map of Forest Crown Density (1996-97 and 2000-01)

References

Ardo, J. 1992. Volume quantification of coniferous forest compartments using spectral radiance recorded by Landsat Thematic Mapper, *Int. J. Remote Sens.*, 13(9), pp. 1779-1786.

Champion, H.G., and Seth, S.K. 1968. A Revised Survey of The Forest Types of India, Govt. of India Press, New Delhi.

Franklin, J. and Strahler, A.H. 1988. Thematic mapper analysis of coniferous forest structure and composition, *Int. J. Remote Sen.*, 26, pp. 809-825.

Guisan, A., Edwards, T.C. and Hastie, T. 2000. Generalized linear and generalized additive models in studies of species distributions: setting the scene, *Ecological Modelling*, 157(2-3), pp. 89-100.

Jaques, D. and Cihlar, J. 1980. Estimation of vegetation biomass in Bunchgrass rangelands of Alberta using Landsat imagery, *Rem. Sens. Earth Resources*, (Eds. Shahsrokhi, F. and Paludan, T.), III, pp. 227-250.

Oza, M.P., Srivastava, V.K and Devaiah, P.K. 1996. Estimating the tree volume in tropical dry deciduous forest from Landsat TM data, *Geocarto International*, 11(4), pp. 33-39.

Paliwal, Rakesh 2001. Ph.D. Thesis 'Application of Fuzzy Logic based models for Sustainable Development'.

Paliwal, Rakesh 2003. Forest Volume Estimation using Remote Sensing, ISRO Induction Training Program-2003 Report.

RRSSC-J. 2004. NR-Census Mission 'A Prototype Study of Jhalawar Distrcit, Rajasthan'.

CONSTRAINTS IN AGROFORESTRY OF MEDICINAL PLANTS IN RAJASTHAN: SITUATION ANALYSIS AND SUGGESTIVE REMEDIAL MEASURES

P.K. Dam, R.C. Sharma, Vinod Joshi, Manju Singhi and Anil Purohit

Desert Medicine Research Centre, Jodhpur

ABSTRACT

Ecology influences the perception and practice of ethnomedicine in the population. 338 respondents from 11 villags of Jodhpur, Jaisalmer and Nagaur district were covered which comes under Luni, Marusthali and Shekhawati zones, respectively. 300 tribal respondents inhabiting 10 villages of Ghatol tehsil from Banswara district were interviewed considering that percentage distribution of tribal people to total population was highest in Banswara district. 26 traditional healers from tribal area and 16 from desert part were interviewed assuming their authenticity to identify medicinal plants by vernacular name. Their perception were documented concerning habitat, cultivation pattern and their constraints, seasonality and mode of application of each plant to alleviate certain selected health problems. As many as 114 medicaments prepared from 81 plant species, to combat 21 selected health problems were reported by the tribal respondents. Disease wise distribution of these 81 plant species according to their kind of stems (i.e. herb, shrub, climber and tree) and magnitude of utilizing their root portion/whole plant in order to assess the load of wild harvesting has been provided here. Situation demands a prompt intervention by the local authority towards providing a comprehensive guideline to the tribal inhabitants in regulating their wild harvesting and to plant a new sapling/ make propagation by seeds after eradicating a medicinal plant for using it's root or whole plant. The constraints are: intense biotic interference (excessive grazing, heavy lopping, illicit felling, random harvesting and vegetation affected by animals) use of chemical fertilizers & insecticides, leading to fall in sustained productivity for prolonged duration, constant use of irrigation devices to enhance crop production in the field, result into downfall of ground water level throughout the area, lack of awareness (about demand, market value and utilizable portion) and lack of concerted effort of the community to venture for large-scale production of desert medicinal plants on co-operative basis.

Some suggestive messages provided to community to overcome existing degradation are: medicinal plants remain unaffected by drought as compared to crop plants; community should utilize excellent and very good lands for crop production; and good, fair and unproductive land-use for cultivating medicinal plants, medicinal plants like *Withania somnifera*, *Jatropha curcas*, *Commiphora mukul*, *Tribulus terrestris*, *Cassia angustifolia* etc. are not consumed by any predator animal and awareness about techniques of fencing system to overcome the environmental constraints. More stringent regulations need be developed to ensure their adequate growth and proliferation in desert and tribal area.

Further, techniques of multi-tier-agriculture-system concerning cultivation of medicinal plants need to be provided for optimum economic growth.

Introduction

Conservation on International Trade in Endangered species of World Fauna and Flora (CITES) provides an international legal framework to the regulation of trade in those endangered plant and animal species that are exploited commercially. During past three decades India has witnessed a vast development in the sphere of industry which resulted in destruction of nature and natural resources (e.g. Kalu River's methyl-mercury and lead poisoning during 1978-79).

The Department of Environment under direct control of Prime Minister works in consultation with NCEP and NEDB (National Ecosystem Development Board) to implement programme through MAB (Indian National Man and the Biosphere Committee) to study the impact of human interference on natural ecosystem.

The Directive Principle (Article 48 and 51A) of the Indian Constitution has assigned permanent responsibility to the state, as well as, every citizen of India for undertaking improvement of natural environment that includes forest and wild life. Studies revealed the biodiversity of Rajasthan that harbors the enriched variety of medicinal plants (Shetty & Singh, 1993; Bhandari, 1978; Dam, *et al.* 2000; Suresh Kumar & Farzana, 2000 & Trivedi, 2002).

To meet the increased demand of plant medicine of the ever increasing human population in Rajasthan, it is essential that more land is brought under plough and productivity of the existing arable land is maintained on an optimum level.

Culturable wastelands cover a large area in the state (Table-1). Due to inherent limitations, it is not feasible to bring all the area under arable use, but a large chunk have the potential for improvement. An attempt has been made here to explore the possibility of their improvement and management.

The present study is established over the background of setting up of one 'Herbal garden' in DMRC campus during 1999-2000, based on the observation concerning social practice of ethnomedicine as reported by the respondents inhabiting desert and tribal areas of Rajasthan.

Table 1. Different category of wastelands in the state

S.No.	Category	Area (ha)	% of total wasteland area	% of total geographical area of Rajasthan
1	Gullied/Ravines	403,784	4.36	1.71
2	Undulating upland with or without scrub	2,602,297	28.11	7.60
3	Water logged	75,551	0.81	0.22
4	Salt affected	178,716	1.93	0.52
5	Sandy area	4,599,531	46.69	13.43
6	Degraded Forest	948,851	10.26	2.77
7	Barren rocky/Stony Waste/Sheet Rock Area	448,120	4.84	1.30
	Total	9,256,850	100.00	

Source: National Remote Sensing Agency, Hyderabad, National Project of wasteland mapping using IRSLISS 1 satellite data

Objectives

The Thar Desert in itself is a natural ecosystem. The directive principle (Art. 48 and 51 A) of the Indian Constitution provide for full protection of the natural environment (including forests and wildlife). Hence, the present study was undertaken:

- To assess the reasons of degeneration of plants/ animals with ethno-medicinal importance in the Thar desert and Tribal part of Rajasthan.
- To enumerate the environmental constraints for cultivation of medicinal plants.
- To cite certain remedial measures to overcome the constraints for cultivation of medicinal plants in study area.

Materials and Methods

Ecology influences the perception and practice of ethnomedicine in the population. Rajasthan comprising of nine agro-climatic zones harbors a vast variety of medicinal plants. These nine zones are different in terms of 1) water table 2) water salinity 3) geographical location and 4) geological formation.

338 respondents from 11 villages of Jodhpur, Jaisalmer and Nagaur district were covered which comes under Luni, Marusthali and Shekhawati zones, respectively (Figure - 1). In Rajasthan, the Scheduled Tribe population to total population is highest in Banswara (75.4 %). Keeping in mind that an area representing higher concentration of tribal population may provide an ideal representation for any tribal study, 10 villages from Banswara district were selected, at least 25 % of total village population were tribal.

Study Area

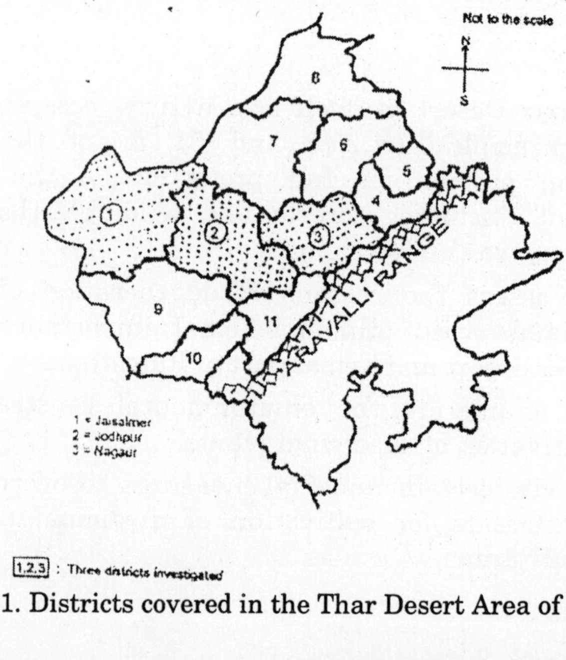

Figure: 1. Districts covered in the Thar Desert Area of Rajasthan.

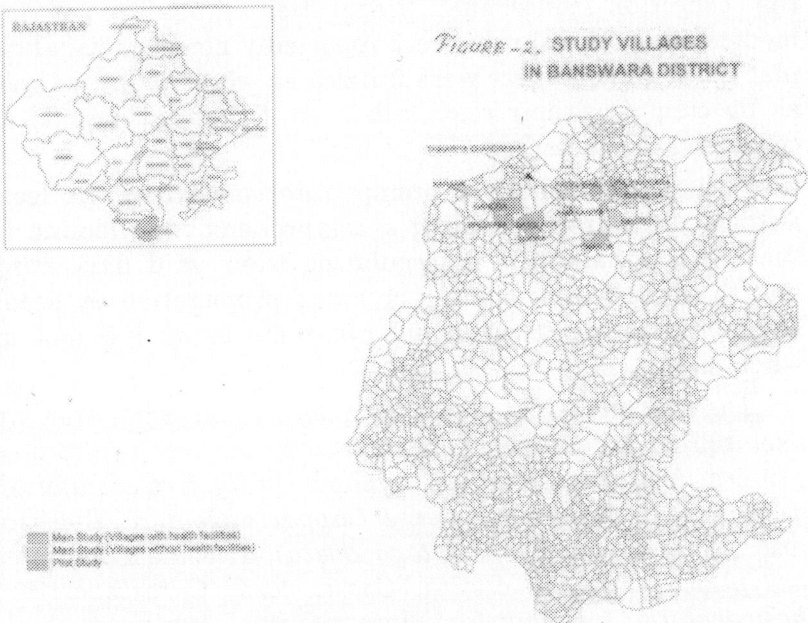

Figure: 2. Study villages in Banswara district, Rajasthan.

300 tribal respondents were interviewed from 10 villages of Ghatol tehsil in Banswara district (Figure 2). Further, 26 traditional healers from tribal area and 16 from desert part were interviewed assuming their authenticity to identify medicinal plants by vernacular name.

Results and Discussions

As many as 114 medicaments prepared from 81 plant species, to combat 21 selected health problems were reported by the tribal respondents. Disease wise distribution of these 81 plant species according to their kind of stems (i.e. herb, shrub, climber, tree) and magnitude of utilizing their root portion/whole plant in order to assess the load of wild harvesting has been provided here (Table 2).

In comparison to trees, the herbs are the most vulnerable group of plants as majority of them were wildly harvested by their roots and whole plants in order to fulfill the condition concerning preparation procedure of the medicament. The shrubs were reportedly not used as whole plant, but the climbers were utilized as whole plant, as well as by chopping their roots, albeit in a reduced degree, as compared to the herbs.

Situation demands a prompt intervention by the local authority towards providing a comprehensive guideline to the tribal inhabitants in regulating their wild harvesting and to plant a new sapling or make propagation by seeds after eradicating a medicinal plant for using it's root or whole plant.

Certain medicinal plants from desert and tribal area are exemplified here which help in remedy of fever (*Calotropis procera, Azadirachta indica*), typhoid (*Indigofera cordifolia*), Toothache (*Solanum surattense, Capparis decidua*), Constipation (*Cassia italica, Plantago ovata*), Eczema (*Balanites aegyptiaca, Fagonia cretica, Cassia tora, Cardiospermum halicacabum, Holoptelea integrifolia*), multiple boils (*Argemone mexicana, Clerodendrum phlomidis*) and malaria *Enicostemma littorale* etc. require immediate conservation efforts against intense biotic pressure.

Respondents were given message to grow medicinal plants bearing least competitive instinct for survival like Jatropa curcas, *Acacia senegal, Withania somnifera, Tinospora cordifolia, Caligonum polygonoides, Tecomella undulata, Prosopis cineraria, Caesalpinia crista* and migratory species like *Andrographis paniculata* (i.e. kalmegh) of family Acanthaceae as well as, *Alocasia indica* (i.e. Arum) of family Araceae. Cultivation techniques, precautionary measurements for minimizing stunted growth of the plants sown in the field and possible economic outcome of each of the aforesaid plants was widely discussed and communicated to the respondents. For example, following beneficiary points to cultivate Jatropha curcas was incorporated to the respondents:

1. No animal ingests any part of *Jatropha curcas.*

2. It requires minimum water for subsistence.
3. In majority, it is a disease free plant having life span of 60-70 years.
4. All leaves falling over the land making a layer of 6-7 inches makes ideal organic fertilizer availability that enhance moisture content retention of the land.
5. In February, plantation to be done; In July, the plant would attain its height up to 6-7 feet, which will in turn gain approximate optimum height of 12 feet during July of next year.
6. Shape of leave of *Jatropha* is such that it can hold rain water which cumulatively drops over the soil to help in checking soil erosion and proper water absorption by soil for the cultivation of plant by releasing water drop by drop through leaves for a prolonged period after it rains.
7. In winter, beneath the *Jatropha* tree, 15°-35° temperature is maintained, which is ideal for maintaining easy flow of water availability for the plant in desert during winter.

Constraints for cultivation of plants (Medicinal)

1. Regular extraction of plants of ethno-medicine by contractors/ Money Lenders- through Villagers
2. Excessive grazing by domestic cattle
3. Occurrence of fire during summer
4. Unscientific way of plucking flowers/ fruits/ twigs stem cutting.
5. By picnic/ hunting/ film shooting
6. Excessive tree felling by timber contractor
7. Flood
8. Draught
9. Irregular monsoon
10. Natural Calamities (Sand Storm/ Heavy rain)

Table 2. Diseasewise distribution of medicinal plants according to kinds of stem and use of their root and whole plant

S. No.	Health Problems	Total no. of medicaments	Total no. of plants involved	Total Number of			Total Number of			Total Number of			Total Number of		
				Herbs	Roots	Whole	Shrubs	Roots	Whole	Climber	Root	Whole	Trees	Root	Whole
1.	Leukorrhea	6	6	1	-	1	-	-	-	1	1	-	4	-	-
2.	Eczema/boils	8	9	2	-	1	1	-	-	2	-	-	4	-	-
3.	Jaundice	6	8	3	1	2	-	-	-	2	-	1	3	-	-
4.	Malaria	3	3	2	-	2	-	-	-	1	-	-	-	-	-
5.	Fever	9	12	4	1	2	3	-	-	1	-	-	4	-	-
6.	Bleeding due to injury	11	7	1	-	-	4	-	-	-	-	-	2	-	-
7.	Snake Bite	5	5	2	2	-	2	1	-	1	1	-	-	-	-
8.	Scorpion Sting	5	5	-	-	-	4	2	-	-	-	-	1	-	-
9.	Prolapse of Uterus	2	4	-	-	-	1	-	-	-	-	-	3	-	-
10.	Excessive bleeding other than M.C.	5	6	1	-	-	-	-	-	-	-	-	5	-	-
11.	Scrofulodesma	4	8	1	1	-	1	-	-	1	1	-	5	1	-
12.	Intestinal Worm	4	4	-	-	-	2	2	-	-	-	-	2	-	-
13.	Diarrhoea	10	13	-	-	-	4	1	-	-	-	-	9	2	-
14.	Constipation	6	4	2	1	-	-	-	-	-	-	-	2	-	-
15.	Headache	4	3	2	-	2	1	-	-	-	-	-	-	-	-
16.	Cough & Cold	4	5	2	-	-	2	-	-	-	-	-	1	-	-
17.	Asthma	2	2	-	-	-	1	-	-	-	-	-	1	-	-
18.	Fracture & Sprain	3	5	-	-	-	-	-	-	2	1	-	3	2	-
19.	Typhoid	6	5	2	1	1	-	-	-	-	-	-	3	-	-
20.	Haemorrhoids	5	4	2	2	-	-	-	-	-	-	-	2	-	-
21.	Indigenous birth control method	6	7	2	1	-	1	1	-	1	-	1	3	-	-
	Pooled	**114**	**125**	**29**	**10**	**11**	**27**	**7**	**-**	**12**	**4**	**2**	**57**	**5**	**-**

11. Wastage by Predator animals (Figure-3).
12. Use of Chemical fertilizers & insecticides, leading to fall in sustained productivity for prolonged duration.
13. Constant use of irrigation devices to enhance crop production in the field resulting into down fall of ground-water level throughout the area.
14. Lack of awareness about demand, market value and utilizable portion of medicinal plants grown in their surroundings.
15. Lack of concerted effort of the community to venture into large-scale production of medicinal plants on co-operative basis.
16. Inadequate availability of indigenously prepared herbal insecticide, which is inevitable for keeping the medicinal plant products as disease free and pesticide free.

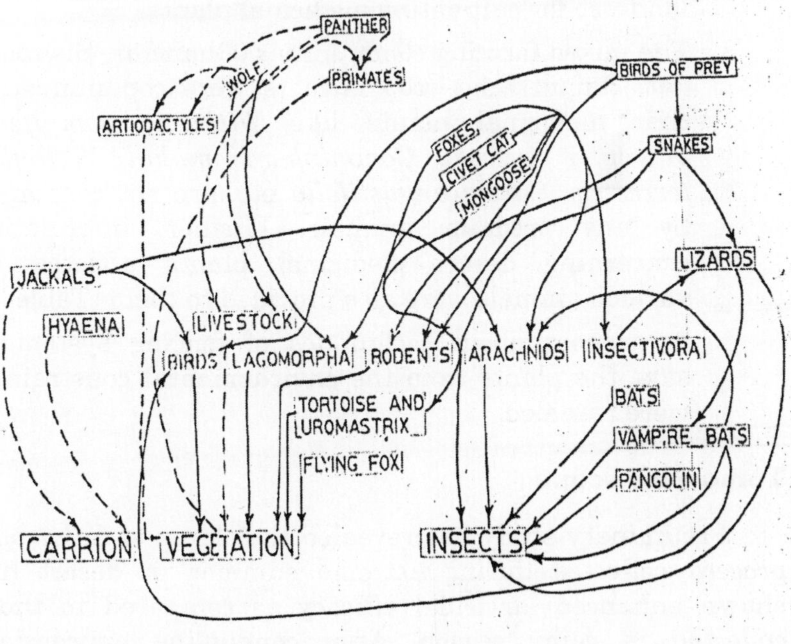

Figure: 3. Food Web of Tharian Mammals (Prakash I. 1975)

Suggestive Messages Provided to Community to overcome Existing Degradation (Through Extension/ IEC Activities)

1. Respondents were communicated that medicinal plants remain unaffected by drought as compared to crop plants. Drought usually affects optimum production of fruits/seeds which are consumable parts of crop plants. Conversely, in case of medicinal plants; root, bark, leaves, latex and sometimes whole plants are utilized to cure illness. Utilizable parts of medicinal plants in desert are capable to withstand desiccation due to drought.
2. Community was given message to utilize excellent (i.e. Badio/Unav) and very good (i.e. kundala/ Jodio) lands for crop production and good (Dabla), fair (Dhal/Tiba) and unproductive (Magrio/ Hills/ Ussar) land-use for cultivating medicinal plants.
3. The mixed farming communities (Ghanchis, Bishnois, Jat, Rajput) and crop farmers were communicated that medicinal plants like *Withania somnifera, Jatropha curcas, Commiphora mukul, Tribulus terrestris, Cassia angustifolia* etc. are not consumed by any predator animals. Further, information concerning desert medicinal plants affected by various animals were also provided to them (Table 3).
4. Awareness about techniques of fencing system to save the plants from the environmental constraints, were revealed.

Tomorrow scope

Preliminary study has revealed that latex of *Calotropis procera* collected during extreme summer in desert has shown enhanced larvicidal efficacy as compared to those collected in rainy season. After conducting appropriate preclinical/clinical studies of some autochthonous medicinal plants from the study area, the status of the indigenously prepared medicaments would be scientifically proven, evidence based and be supported by solid data. Thereafter, the community of the crop farmers need to be educated to grow specific medicinal plants in their field, in addition to crop plants. In desert, there is minimum drought related risk factor on cultivation of medicinal plants.

Further, techniques of multi-tier-agriculture-system concerning cultivation of medicinal plants need to be provided for optimum economic growth.

- The officers and guards employed in the forest should be imparted training as regards other associated group of medicinal plants grown in that forest.
- Policy for conservation of forest products like medicinal plants need to be determined before undertaking any programme for cultivation of medicinal plants.

Table 3. Desert vegetation affected by various animals

Plant Species	Part Consumed	Predator Animals
Prosopis cineraria	Sapling	Lagomorphs, i.e. *Lepus nigricollis*
(Khejari)	Sapling	Lagomorphs, i.e. *Lepus nigricollis*
Albizzia lebbek (Sares)	Ripe Fruit	Atriodactyles i.e. *Gazella g. bennatti*
Capparis decidua (Ker)	Dry leaves	Atriodactyles i.e. *Gazella g. bennatti*
Calotropis procera (Aak)	Seeds, flowers	Desert Rodents i.e. *Tatera indica*
Calligonum polygonoides (Phog)	Seeds, flowers	Desert Rodents i.e. *Tatera indica*
Zizyphus nummularia (Borti)	Whole Plant	*Locust i.e. Schistocerca gregaria*
Euphorbia granulata (Dudheli)	Bark, Leaves	*Locust i.e. Schistocerca gregaria*
Capparis decidua,	Bark, leaves	----*do*-----
Prosopis cineraria,	Roots	----*do*-----
Wheat, Millet, Moth, Makai	Roots	Termites i.e. *Microcerotermes raja*
Pluchea lanceolata (Rasna)	Scrub	-----do-----
	Leaves	*Ardeotis nigriceps*
	Pods, leaves	*Macaca mulatta,*

Commiphora wightii (Guggul)	*Presbytis entellus*
	Sheeps & Goats
Capparis, Euphorbia	
All Native Vegetation	
Acacia nilotica (Babul)	

- Conservation of improved and disease-free variety of medicinal plants in adequate number should be the first step before initiating their agro-forestry.
- Certain medicinal plants grow as dominant plant variety in a particular forest set up, which need to be earmarked for promoting their optimum growth, as well as, registering them for improved mode of marketing and distribution system.
- Endangered species in a particular forest need to be identified first. Thereafter, the tribal respondents to be taken into confidence for sowing these plants of endangered sp. by propagation of the seeds, as and when they harvest the plant for preparing herbal medicine.
- Booklets in Marwari and Mewari language need to be published as regards complete guide book for cultivation of indigenous medicinal plants so that crop farmers and mixed farming communities may improve their awareness of medicinal plants, with special attention to (1) Climate (2) Temperature (3) Pollution free environment (4) Appropriate soil condition (5) Selection of appropriate plants to be sown along with main plants (6) Technique of cultivation etc.

Conclusion

As ethnomedicine treatment was less costly, less technologically oriented, more accessible and equally effective with no side effects, thereby, respondents expressed their satisfaction after using ethnomedicine compared to allopathic treatments, although they were ignorant about the active principles contained by the plants responsible for curing those health problems. Under such circumstances, it is the need of the hour, to intervene the communities so that they can initiate a sustainable harvesting of medicinal plants for treatments and health promotion.

Acknowledgement

We would like to express our thanks to Dr. Himmat Singh, Research Assistant, for his meticulous secretarial support in finalizing the manuscript.

References

Shetty, B.V. and Singh, V. 1993. "Flora of Rajasthan" Vol. 3. Calcutta: Botanical Survey of India, : 861-1245.

Bhandari, M.M. 1978. Flora of the Indian Desert. Jodhpur: MPS Repros, : 1-435.

Dam, P.K.; Ramnath, T.; Yadav, S.P. and Tyagi, B.K. 2000. "Use of Ethnomedicinal practices in the Thar Desert, North Western India." *Journal of Economic and Taxonomic Botany*, 24 (2): 303-318.

Kumar, Suresh and Praveen Farzana. 2000. "Floristic diversity as a source of household, traditional and commercialised remedies in arid western Rajasthan". *Journal of Economic and Taxonomic Botany*, 24 (2): 495-505.

Trivedi, P.C. 2002. Ethno-medicinal Plants of Rajasthan State, India in: Trivedi, P.C. (Ed.). Ethnobotany. Jaipur, India: Aaviskar Publishers, Distributors, : 412-439.

Prakash, I. 1975. The Ecology and Zoo Geography of Mammals, In: Gupta R. & Prakash I. eds. Environmental analysis of the Thar desert. Dehra Dun English Book Depot, : 448-465.

Bhatia, D.; Shahi, V.K.; Ruhii, S.D. and Sharma, K.D. 1995. "Observations on the role of Ayush-64 in the malaria epidemic in Western Rajasthan". Seminar on Research in Ayurveda and Siddha, CCRAS. New Delhi, 20-22 March, 1995.

SUCCESSFUL DEMONSTRATION OF URBAN FORESTRY MODEL IN JODHPUR CITY AND STRATEGIES FOR ITS EXTENSION IN OTHER URBAN AREAS OF ARID REGION

Pradeep Chaudhry, Arvind Apte and R.L. Srivastava
Arid Forest Research Institute, New Pali Road Jodhpur-342 005.

ABSTRACT

Trees and green spaces play an important role in providing improved living conditions in the cities. Though, trees have played an important role in human settlements throughout history, their full value to urban-dwellers has been recognized recently. Based on future projections, the new millennium may rightly be called as urban millennium. By the year 2030, almost 85% of Latin Americans and 50% of Afro-Asians will live in cities. Urban areas in developing countries have multi-faceted problems like paucity of safe water, inadequate waste management, pollution hazards and degradation of sensitive lands. Multi-resource urban forest management is one of the solutions to these problems. For example, in Durban, South Africa, multifunctional parks are a component of slum improvement programmes; parks are used for storm water catchments and wastewater, sewage treatment, recreation and gardening. In an arid zone city environment like Jodhpur, developing urban greenery is really a challenging task in view of harsh conditions like low and erratic precipitation, frequent drought, intense radiation, low humidity, high evapo-transpiration, poor soil conditions and poor vegetation cover.

Arid Forest Research Institute at the request of district administration, Jodhpur took up the work of developing urban forestry models on different road sites of Jodhpur city from the year 2001-2002. Till the year 2004-2005, about 6000 plants of different species like *Dalbergia sissoo, Azadirachta indica, Pongamia pinnata, Cassia siamea, Cassia fistula, Alastonia scholaris, Tecomella undulata, Bauhinia variegata* and *Delonix regia* were planted and maintained over a length of about 6.50 km. Results have been very satisfactory with survival percentage more than ninety percent. At the age of 51 months, average height recorded in respect of *Dalbergia sissoo* is 883 cm, *Azadirachta indica* as 681 cm and *Cassia siamea* as 539 cm. complete failure was recorded in respect of survival of species like *Delonix regia* and *Bauhinia variegata.*

Proper strategies and planning have to be mixed together for development of urban greenery in big cities right from initiation of town planning exercise. Responsibilities of implementing departments/agencies and proper coordination amongst themselves are the key for successful urban forestry programmes. Big industrial houses, newspaper groups, automobile dealers etc. have to be properly involved in maintenance of urban greenery. Annual competition for best developed lawns, combination of shrubs, herbs and flowers in such green areas; among schools, colleges, housing societies and individual house gardens will help in developing green attitude among citizens. Partially treated sewage water should not go waste and has to be utilized for betterment of public parks and gardens. A detailed study is needed to quantify the economic and social benefits of urban forestry to affect resource-allocation and to develop a scientific knowledge base in respect of city's urban greenery. In India, one of such kind of study was done for Chandigarh city, where annual recreational use value of city's urban forestry assets was estimated as Rs. 27.50 millions from the point of view of city's residents. Mean willingness to pay for betterment of existing landscape features in green belts and for creating new ones, was estimated as Rs. 153/- per family per year for a period of five years. Institutes like CAZRI and AFRI have to play a major role in not only such kind of studies but also in developing quality planting stock with quality seeds and good clonal material for planting and establishment of seedlings in arid zone urban environment.

INTRODUCTION

Trees have played an important role in human settlements throughout history. However, their full value to urban-dwellers has recently been recognized. Based on the future projections, the new millennium may rightly be termed as urban millennium. Urban areas in developing countries will account for nearly 90 percent of projected world population increase of 2700 million people between 1995 and 2030. By the year 2030, almost 85 percent of Latin Americans and 50 percent of Afro-Asians will live in cities. The most explosive urban growth is expected in Asia and Africa. Asia will have the largest urban population in the world, with almost twice as many people living in cities as in Africa and Latin America (UN, 1998). As the bulk of world's population shifts from rural to urban areas, poverty is becoming an increasingly urban phenomenon. As per the World Bank estimates, within the next 20 years more poor people will live in the cities than in rural areas (WRI, 1996).

Urban areas in developing countries have multi-faceted problems like paucity of safe water, inadequate waste management, pollution hazards, occupation and degradation of sensitive lands, flooding and soil erosion in unauthorized settlements. Seeland (2005) mentions following standard problems related to public green space management in urban areas of developing countries:

1. Increasing population rate/over population.
2. Migration from rural areas to the large agglomerations.
3. Air (toxic gases, heat and dust) and water pollution.
4. Little problem awareness in the public.
5. Little/no support by politicians concerning green space and related matters.

6. Solid waste disposal problems.
7. Lack of hygiene and sanitation.
8. Poverty creates slum and squatter areas.
9. Lack of open free space in and around cities.
10. Encroachment of wastelands by in-migrating people from rural areas, mostly with herds of domestic animals.

Multi resource urban forest management is one of the solutions to some of these problems. For example, in Durban, South Africa, multifunctional parks are a component of slum improvement programmes; parks are used for storm water catchments and wastewater, sewage treatment, recreation and gardening (Kuchelmeister, 2000).

Urban forestry and its benefits

Many urban foresters in industrialized countries use the terms "urban greening" and "urban forestry" interchangeably (Miller, 1997). In broader terms, urban forests are regarded as the entire forest area influenced by the urban population. In a more restricted sense, urban forestry relates to trees and woodlands and emerging woodlands on vacant and derelict land (Kuchelmeister, 2000). Urban forestry is a branch of forestry with the objective of cultivation and management of trees for their present potential contributions to the physiological, sociological and economic well-being of urban society. The literature indicates that the research in urban forestry has been advancing rapidly in North America and Europe through concerted actions and significant resource allocation. The list of goods and services that urban forestry can provide is impressive. Trees and green spaces help keep cities cool, act as natural filters and noise absorbers; improve micro-climates and the quality of natural resources including soil, water, vegetation and wildlife. Trees contribute significantly to the aesthetic beauty of cities, thereby helping to maintain the psychological health of the inhabitants. The tangible benefits of urban forests include availability of fruit, fuel wood and small timber, whereas, intangible benefits are much more than tangible benefits.

Material and methods

Development of urban forestry models in harsh arid conditions like low and erratic precipitation, frequent drought, intense radiation, low humidity, high evapo-transpiration, and poor soil conditions in Thar Desert, which is also facing maximum biotic pressure among all the World's deserts, is really a challenging task.

Arid Forest Research Institute, Jodhpur at the request of district administration, Jodhpur took up the work of developing urban forestry models on different road sites of Jodhpur city from the year 2001-2002. Till the year 2004-2005, about 6000 plants of different species like *Dalbergia sissoo, Azadirachta indica, Pongamia pinnata, Cassia siamea, Cassia fistula, Alastonia scholaris, Tecomella undulata, Bauhinia variegata* and *Delonix regia* were planted and maintained over a length of about 6.50 km. in the city. Results have been very encouraging with survival percentage more than ninety percent. Effective barbed wire protection, sticking to correct watering schedule and protection of plants from termites etc. are the main reasons for good survival percentage. Wherever possible, sewage water was utilized for watering of plants.

RESULT AND DISCUSSION

At the age of 51 months, average height recorded in respect of *Dalbergia sissoo* was 883 cm, *Azadirachta indica* as 681 cm and *Cassia siamea* as 539 cm. Complete failure was recorded in respect of survival of species like *Delonix regia* and *Bauhinia variegata.* Below given table gives overall picture regarding performance of different species:

Species	Average collar girth (cm)	Average height (cm)
Dalbergia sissoo	63	883
Azadirachta indica	54	681
Cassia siamea	39	539
Alastonia scholaris	40	389
Pongamia pinnata	34	386
Cassia fistula	38	385
Tecomella undulata	27	377

Strategies to be adopted for extension

Proper planning is the key to success of any management task and artificial plantation development and management is no exception to it. Generally, there are many government funded departments, corporations, trusts etc in big cities who can be involved in urban forestry works but their correct identification based on their available resources, strengths/weaknesses and their coordination with other agencies for plantation job is very essential for successful development of urban forestry models. It has been seen that after establishment of such models, maintenance is generally neglected. Following the examples of big cities where urban forestry has been developed in India like Bangalore and Chandigarh, the maintenance of green areas must be assigned to big industrial houses, newspaper groups and automobile dealers etc. who can maintain such spots and display proudly their small sized sign-boards for publicity.

Annual competitions among different schools, colleges, housing societies etc. for best maintained parks/gardens in their respective locations must be organized by the district administration or municipal corporations/councils once a year. This will go a long way in developing green attitude among citizens especially children. Town planners of the cities must ensure that partially treated sewage water must be utilized properly in parks/gardens and roadside plantations. This will ensure optimum water utilization in arid region, where water is a precious commodity.

It has been found in some studies conducted outside India that people of the city will not mind paying a nominal entrance fee to well-maintained parks or gardens and such kind of optimum entrance fees have been estimated for different protected areas, urban forests and other areas of scenic beauty by methods prescribed in the literature of environmental-economics. In our country, for urban forestry/greenery, one such kind of study has been conducted in Chandigarh city, where annual recreational use value of city's urban forestry assets was estimated as Rs. 27.50 millions from the point of view of city's residents. Mean willingness to pay (WTP) for betterment of existing landscape features in green belts and for creating new ones, was estimated as Rs. 153/- per family per year for a period of five years (Chaudhry, 2004). Such kinds of studies are needed to quantify the economic and social benefits of urban forestry for effective resource-allocation and to develop a scientific knowledge base in respect of a city's urban greenery. The various benefits, both the market and non-market, of the urban forests are often ignored or grossly under estimated. On the other hand, various developmental options to the area may be attractive to politicians and bureaucrats due to immediate returns from development projects. It has been observed that urban parks and open spaces left for green belt formation in future are subject to development pressure, especially in developing countries. One possible reason for this is that the planners and researchers have been unable to articulate their value in economic terms (More *et al.*, 1998). If benefits from the urban forest areas are not properly assessed, the authorities' decision may be biased to implement development activities such as construction of schools, shopping-complex, housing-flats etc. in open and green spaces. Once the monetary valuation of recreational and other non-market benefits is done, it can be introduced into public decision-making and cost-benefit analysis of the projects.

Various scientific and research institutes in the field of arid forestry like Arid Forest Research Institute, Jodhpur; Central Arid Zone Research Institute, Jodhpur and related universities of this region have to play a major role in conducting such kind of studies in addition to development of quality planting stock for proper and easy establishment of seedlings in the arid zone urban environment.

References

Chaudhry, P. 2004. Valuing recreational benefits of urban forestry- A case study of Chandigarh city. *Ph.D. thesis.* (Submitted) FRI (Deemed university), Dehradun.p. 123.

Kuchelmeister, G. 2000. Trees for the urban millennium: urban forestry update. *Unasylva no. 200* (special issue: Trees outside forests), 51: 2000/2001.

Miller, R. W. 1997. Urban Forestry: Planning and managing green spaces. 2nd edition. Prentice Hall Inc., Upper Saddle River, New Jersey, p. 502.

More, T. A., Stevens, T. and Allen, P. G. 1998. Valuation of urban parks. *Landscape and urban planning*, 15(1-2): 139-152.

Seeland, K. 2005. Trees outside forests in an Indian urban environment. In: "Multipurpose trees in the Tropics: Management and Improvement Strategies" (Eds.) V.P. Tewari and R.L. Srivastava. Proceedings of IUFRO sponsored international workshop held at Arid Forest Research Institute, Jodhpur (22-25th Nov, 2004) pp. 486-492. Scientific Publishers, Jodhpur.

U.N. 1998. World Urbanization prospects: the 1996 revision. United Nations, New York.

W.R.I. 1996. World Resources 1996-1997: The Urban Environment. World Resources Institute, Washington DC, U.S.A. p. 400-8.

ROLE OF FOREST PROTECTION RESEARCH IN THE EXTENSION ACTIVITIES

S.I. Ahmed and K.K. Srivastava
Arid Forest Research Institute, Jodhpur-342 005.

ABSTRACT

The extension research related to the forest protection division includes the studies on management of Khejri (*Prosopis cineraria*) mortality in Rajasthan. Problem of khejri mortality is related to biotic and abiotic factors. Khejri decline in farmer's fields and mixed Khejri forests is a complex phenomenon involving environmental stressors, often drought, indiscriminate lopping practice, root diseases and insect pests particularly wood boring insects. The paper also contains the information on integrated pest management of rohida and neem. VAM association and VAM production work has also been as one of the important aspects of the extension activities of the division.

Introduction

Insects and diseases exert negative impact on forest trees in several ways. They can kill trees; reduce their growth; degrade wood and other products; cause dieback, decline and deformity change the composition of the forest; reduce biological diversity; affect water quality and quantity; create safety hazards; increase fire risk; reduce the quality of the landscape and cause other kinds of damage.

Some of these types of damage may not be significant if they are not detrimental to the intended use of the forests. It is important to note that pest outbreaks do not respect ownership boundaries. While the management strategies discussed below may lead to a measure of protection of forest lands from destructive insect or disease activity, failure of a landowner or land-manager to control pest outbreaks can affect other owners lands. Passive management of important tree species such as khejri, rohida and neem in arid and semi-arid regions of Rajasthan can easily lead to pest population spillover and negatively affect forest resources of adjacent landowners.

1. Management of Khejri mortality in Rajasthan

The problem is related to biotic and abiotic factors. Khejri decline in farmer's fields and mixed Khejri forests is a disease complex involving environmental stressors, often drought, root diseases and insect pests particularly wood boring insects. Large-scale Khejri mortality in north western Rajasthan is being faced due to successive and indiscriminate lopping practice followed by secondary infestations of shoot and root borers and severe infections of plant pathogenic diseases.

Factors responsible for mortality

Biotic Factors

In the recent studies, conducted by the Arid Forest Research Institute, Jodhpur in order to find out the actual causes/factors responsible for the large scale drying and the subsequent die-back of mature Khejri trees in western districts of Rajasthan, it is closely examined that this devastating problem has primarily been raised with the cumulative effects of indiscriminate and successive lopping followed by a secondary infestation of shoot borers, *Derolus iranensis*, *Aeolesthes holosericea*, *Hypoeschrus indicus* and *Acanthophorus serraticornis*. The borers lay their eggs on the freshly lopped top portions of shoots, branches main trunk and root portions of the infected trees and complete their life-cycles by destroying the plant tissues of the main branches, stems and root system. The larvae excavate long, irregular tunnels in the sap and heartwood of the main trunk and roots. The borer attack is followed by a tertiary infection of fungus diseases. The infected samples reveal the presence of three highly infective species of *Fungi imperfactii viz., Alternaria alternata*; *Phoma* sp. and *Botryodiplodia theobromae* which cause the die-back disease in mature trees of Khejri as a result of which the tree starts drying from the top.

D. iranensis, A. holoserecea, H. indicus and *A. serraticornis* have also been recorded for the first time on *Prosopis cineraria* from arid areas of Rajasthan. Though, Gahen (1906) has described the systematic position of *D. descicollis* (*iranensis*) and reported this species to have been collected from Karachi during 1906 but nothing has been mentioned by him on its host plants. Whereas, Beeson (1941) reported *D. iranensis* as a pest of *Acacia modesta* and *Heritiera fomes*. Ahmed *et al.*, (2001) have described the bio-ecology and control of *D. iranensis* based on the field observations made during the studies, conducted in the affected areas of Rajasthan.

Abiotic Factors

Data collected on abiotic factors such as repeated and indiscriminate lopping of Khejri trees, continuous depletion of water tables, over exploitation of ground water through tube wells, effect of low rainfall and change in soil properties and agricultural practices from 140 different localities in four districts of north western Rajasthan *viz.*, Nagaur, Sikar, Churu and Jhunjhunu indicate their positive involvement in large-scale drying of Khejri in western zone of Rajasthan. Further studies are being carried out on the specific role of the most potential abiotic factors in the large-scale decline of Khejri trees in Rajasthan.

Management strategy

It has been observed that the Khejri trees which were lopped after a gap of one year, with one third canopy unlopped, showed lesser incidence of pest and disease infestations whereas the trees, with complete lopping without any gap, exhibited high degree of infections of disease and insect borers. Hence, in order to protect the trees from further spread of pest and disease infestations, following remedial measures are suggested:

The drying trees can be managed and cured by application of root and shoot treatment simultaneously as recommended below:

- A gap of one year for khejri lopping is advisable in order to recover the tree from the injuries made by pests and diseases.
- Soil around the root system should be loosened and treated with a suspension containing Bavistin (0.1%) + Chloropyriphos (0.1%) + leader or Agro-min (2ml/lit) @ 5 lt. aqueous suspension per tree.
- The lopped portions / open wounds should be treated with AFRI PASTE* to check the fungal infection and egg laying by the shoot borers.
- Severely infected and dried trees should be uprooted and the felled trees should be lifted from the vicinity immediately after their exploitation.
- The trees, showing partial die-back symptoms should be dealt for their treatment on priority in order to check further spread of pest / disease.

Preparation of AFRI paste: *Mix Copper Carbonate (1 part) and Red Lead (1 part) homogeneously then add white petroleum jelly or raw linseed oil (*Alsi ka tel*) (2 parts) and add insecticide (Monocrotophos) (@ 3 ml/Kg) (Modified Chaubatia paste) and apply by brush or by hand using gloves. The paste has toxic effect; hence hand should be thoroughly washed with dettol soap after given treatment.

Use of media and press for dissemination of management package

A scientific documentary film *"Khejri Ek Jeevan Rekha-Astitva Ka Sankat"*, has been developed by Arid Forest Research Institute, Jodhpur in collaboration with Electronic Media Research Centre, Jodhpur (EMRC) on the mortality of Khejri (*Prosopis cineraria*) in north-western districts of Rajasthan. The film was telecasted for the first time on 22nd Jan 2003 then further twice on the National TV channel (DD1) of Doordarshan under University Grant Commission programme to apprise the local farmers, NGOs, and plant growers with the causes of Khejri mortality and its management.

2. IPM of gall inducing pests of khejri (*Prosopis cineraria*)

Types of galls and responsible agents on *Prosopis cineraria*

- **Stem Gall -** is caused by *Eurytoma settitibia* Gahan (Eurytomidae: Hymenoptera). **Rachis Gall-** is caused by *Contarinia prosopidis* Mani (Cecidomyiidae: Diptera). **Leaf Gall-** is caused by *Eriophyes prosopidis* Saxena (Eriophydae: Acarina). **Inflorescence Gall**-is caused by *Eriophyes prosopidis* Saxena (Eriophydae: Acarina).

Silvicultural control of gall inducing insects and mites

- In lopped trees (once in three years) of Khejri, the % of gall formation per inflorescence was observed minimum (5.6%) and pod production was recorded as high as 13.3% whereas in unlopped trees, the % of gall formation was maximum (49.5%) and resultant pod production was as minimum as 3.37%.

- The higher production of pods and lesser formation of galls in lopped trees was because of the minimum infestation of gall mite, *E. prosopidis*.

Management

Hence, one-third lopping practice once in three years is recommended to save the tree from infestation of gall insects and mites.

3. IPM of rohida defoliator, *Patialus tecomella*

Patialus tecomella Pajni, Kumar and Rose has been reported as a serious pest of Marwar teak, *Tecomella undulata* (Sm.) Seem (Bignoniaceae), throughout the north-west Rajasthan. This species has been reported as a serious pest of *T. undulata* from Rajasthan and Punjab. Following IPM strategies have been found effective in controlling the pest.

Management

Application of a microbial pesticide, *Beauveria bassiana* (Bal.) has been found effective in managing the population of *Patialus tecomella.* All the weevils were killed after 11 days of application. Hence, recommended for bio-control.

- Mass release of a potential pupal parasite, *Billeae atkinsoni* recorded to keep the pest populations below the level of economic threshold.
- Dhatura (*Datura metal)* leaf extract is recommended to control the pest population.

Foliar spray using monocrotophos (0.02%) was found to be most effective in the month of July.

4. Integrated management of neem insects & diseases

Sap-sucking and defoliating insects

Helopeltis antonii, Pulvinaria maxima, Aonidiella orientalis, Dialeurodes almatus and *Heliothripes hemorr-*

hodalis are the important sap sucking insect pests of neem whereas *Boramia variegata, Eurema* sp., *Myllocerus tenuicornis, Myllocerus laetivirens* and *Achaea janata* are the major defoliating insects causing damage to neem trees.

Management

The above mentioned insect pests can be controlled by foliar application of systemic insecticides i.e., Monocrotophos or Dimethoate 0.02% in combination with Bavistin (0.02%) in water.

Neem shoot borer

- *Laspereysia koenigana* is a serious pest of neem in young plantation in Gujarat. The moths lay eggs on the tender shoots. The larva after hatching starts boring into the shoot and continue throughout its development.

Management

Spraying of Rogor 30 EC @ 0.02% in combination with Blitox @ 0.05% and Vipul @ 2 ml/Lt. has been found very effective in controlling the pests during July to September.

Root feeders of neem

- Termites and white grubs reduce the growth potential of young plants and result in complete cessation of growth and mortality of weak plants.
- *Odontotermes obesus O. gurdaspurensis O. redemanni* and *Holotricha consanguinea* are the main root feeding insects.

Management

Planting pits should be treated with Chloropyriphos @ 0.02% before tree plantation. The treatment operations should be done for atleast two years at six monthly intervals.

5. VAM studies and production work

VAM association in five tree species namely, *A. nilotica*, *A. indica*, *Prosopis cineraria*, *Tecomella undulata* and *Dalbergia sissoo* were carried out in agroforestry areas of Rajasthan and irrigated plantations raised at IGNP areas under the project entitled *"Irrigated water management for tree species"* Sub Project *"Studies on VAM association in irrigated plantations and agroforestry systems* under WB (FREE) project. A total of 36 different VAM fungi belonging to five genera (*Glomus*-26; 3 each of *Gigaspora* and *Sclerocystis* and 2 each of *Acaulospora* and *Scutellospora*) were isolated and identified. Among them *Glomus* was found most dominant genera and *G. fasciculatum* was found most predominant species in the entire tree species investigated. Maximum VAM colonization and number of VAM spore population were recorded in 2nd quarter *i.e.*, July to September (Table 1) in arid areas of Rajasthan. VAM population ranged between 262-436/100 gm rhizosphere soil. Whereas, 302-628 /100 gm soil in IGNP area (Table 1 & 2). Percentage of root colonization was also noticed maximum in July-September and minimum in winter (January-March). VAM inoculated seedlings performed better in all parameters and enhanced growth up to 55 percent as compare to control (Table 4 & 5).

A detailed study was under taken under the project entitled on *"Studies on improving tree productivity of P. cineraria* (Khejri) *through VAM/other biofertilizers"*. The maximum number (36) of AM fungal species were identified from Jodhpur whereas only 13 species were recorded from Jaisalmer. Maximum spore population was recorded after first rain *i.e*, in July–September and minimum in summers. *Glomus fasciculatum* and *G. aggregatum* were found dominant genera while *Scutellospora* and *Sclerocystis species* were less in number. Seedlings inoculated with VAM + Rhizobium were performed better enhanced the growth up to 1.8 times more than control (un-inoculated). Under the Project "Identification of mycorrhizal and Rhizobial association, establishing gene bank and technology transfer to farmers field" funded by Gujarat State Biotechnology Mission, nine species namely *Ailanthus excelsa, Azadirachta indica, Prosopis cineraria, Commiphora mukul, Eucalyptus* sp. *Salvadora* sp, *Acacia catechu Acacia senegal* and *Sapindus indica* were selected. Identification work of VAM is in progress. VAM multiplication work has been started at Basan nursery, Gandhinagar. Under the project "Training on Biofertilizer" was organized at TRC, Gandhinagar.

Table 1. Percentage of Root colonization and VAM spore population of various arid tree species in natural plantations

Tree Species	April - June		July - September		October - December		January - March	
	(%) Root Colonization	No. of VAM Spores	(%) Root Colonization	No. of VAM Spores	(%) Root Colonization	No. of VAM Spores	(%) Root Colonization	No. of VAM Spores
Acacia nilotica	36	142	52	262	44	234	39	159
Acacia tortilis	62	301	70	321	60	310	45	258
Prosopis cineraria	58	289	68	387	64	299	52	271
Eucalyptus camaldulensis	66	366	74	436	62	358	54	294
Tecomella undulata	39	210	62	301	51	218	38	188

Table 2. VAM spore population in various tree species at IGNP area (8 years old plantations)

Tree species	April - June		July - September		October-December		January - March	
	(%) Root Coloni-zation	No. of VAM Spo-res	(%) Root Coloni-zation	No. of VAM Spo-res	(%) Root Coloni-zation	No. of VAM Spo-res	(%) Root Coloni-zation	No. of VAM Spo-res
Acacia nilotica	54	133	69	472	57	749	41	294
Acacia tortilis	78	358	82	628	69	1011	61	374
Dalbergia sissoo	80	363	94	543	82	794	66	404
Eucalyptus camaldulensis	86	339	86	603	88	802	70	426
Tecomella undulata	59	198	72	302	61	431	48	264

Table 3. VAM spore population in various tree species at Agroforestry area.

Tree species	April - June		July - September		October - December		January - March	
	(%) Root Colon-ization	No. of VAM Spo-res	(%) Root Colon-ization	No. of VAM Spo-res	(%) Root Colon-ization	No. of VAM Spo-res	(%) Root Colon-ization	No. of VAM Spo-res
Acacia nilotica	36	142	52	262	44	234	30	159
Acacia tortilis	62	301	70	21	60	310	48	258
Prosopis cineraria	58	289	68	387	64	299	52	271
Eucalyptus camaldulensis	66	366	74	436	62	358	54	294
Tecomella undulata	39	210	62	301	51	218	38	188

Table 4. Effect of VAM inoculation on shoot height, root length & (%) VAM Colonization on various tree species of arid zone.

Tree Species	Mean shoot height (cm)		Mean root length (cm)		(%) VAM Colonization	
	Un - inoculated	Inoculated	Un - inoculated	Inoculated	Un - inoculated	Inoculated
Azadirachta indica	29.2	40.2	8.2	9.3	45	94
Tecomella undulata	34.6	42.8	7.7	10.2	35	90
Ailanthus excelsa	43.4	49.2	10.1	12.4	32	80
Dalbergia sissoo	36.4	42.1	8.1	10.4	36	86

Table 5. Effect of VAM inoculation on biomass and (%) VAM Colonization on different semi-arid tree species (after one year old seedlings)

Tree species	Average biomass (gm)			(%) VAM Colonization		
	UI	I	Increase (%)	UI	I	Increase (%)
Cordia myxa	4.4	5.3	4.77	20	42	52.3
Artocarpus integrifolia	4.0	4.8	16.60	16	36	55.5
Dalbergia sissoo	4.8	6.0	20.00	32	58	44.8
Pongamia pinnata	4.2	5.1	17.60	28	46	39.1
Mangifera indica	5.1	6.2	17.70	30	42	28.6
Alstonia sp.	12.8	18.6	31.10	42	65	35.4

Other Extension activities under VAM technology

- VAM production work was done at TRC, Gandhinagar, and Gujarat. Staff of State Forest Department, Gujarat, Farmers, NGO, s and Forest officials were apprised with the production methodology.
- Training on VAM technology /biofertilizer.

Literature consulted

Ahmed, S.I.; Srivastava, K.K.; Singh, G. and Prasad, R. 2001. A report on the scientific approach to study the causes of mortality of *Prosopis cineraria* (L) Druce (Khejri) in Western Rajasthan. *Report submitted to State Forest Department Rajasthan and DDG Research*, ICFRE, Dehradun.

Ahmed, S.I., Kumar, Shivesh and Paunikar, S.D. 2000. Biological control of *Streblote siva* through NPV and natural enemy complex of insect pests of *Prosopis cineraria* in Rajasthan. *Indian Journal of Forestry,* 23(3): 305-311.

Beeson, C.F.C. 1941. The ecology and control of the forest insects of India and neighbouring countries. Vasant Press, Dehra Dun, India: 1007 pp.

Bhasin, G.P. and Roonwal, M.L. 1958. A list of insect pests of forest plants in India and adjacent countries - *Indian For. Bull.Entomol.* 171: 93 pp.

Brucher, G.E. 1964. The regulation and control of insects by fungi. *Annals of the Entomological Society of Quebec*, 9: 30-42.

Chandra, H and Chandra, S. 1987. *Omanocoris versicolor* (Coreidae : Heteroptera), a new pest on Khejari (*Prosopis spicigera*). *Plant Protection Bulletin.* 37 (3, 4): 39.

Cunningham, J.C. 1985. Status of viruses as biological agents for spruce bud worms. *Proc. Symp. Microbial Control of Spruce Budworms and Gypsy moths.* Windsor Locks, CT, USDA-GTR 100: 175 pp.

Dresner, E. 1949. Culture and use of entomogenous fungi for the control of insect pests. *Contribution of the Boyce Thompson Institute*, 15: 319-335.

ESC 1986. Microbial insecticide in Canada: their registration and use in agriculture, forestry and public and animal health. A report prepared by the special committee of the science policy committee. *Entomological Society of Canada*: 1-43 pp.

Gahen, C.J. 1906. *Fauna of British India* (Coleoptera), Vol. I (Cerambycidae). Today and Tomorrow's Printers and Publishers, Karol Bagh, New Delhi: 135-136 pp.

Gaur, Meeta 1998. Investigations on the morphology bionomics, ecology and management of Rohida defoliator *Patiallus tecomella* Pajni Kumar & Rose (Coleoptera: Curculionidae). *Ph. D Thesis, FRI (Deemed Univ.)* Dehra Dun.

Horn, D.J. 1971. The relationship between parasite, *Tetrasticus incerus* (Hymenoptera: Eulophidae), and its host, the alfalfa weevil, *Hypera postica* (Coleoptera: Curculionidae), in New York. *Canadian Entomologist.*, 103: 83 - 94.

Kumar V.S.K. and Tiwari, V.P. 2000. Effect of lopping on the top feed production and growth of Prosopis cineraria. *Bioresource Technology*, 74: 165 - 168.

Kumar, Shivesh 1999. Bioecology and management of insect pests of *Prosopis* spp., with special emphasis to gall forming insects in Indian Thar desert. *Ph. D Thesis, FRI (Deemed Univ.)* Dehra Dun.

Mathur, R.N. and Singh, B. 1960. A list of insect pests of forest plants in India and the adjacent countries. *Indian Forest Bulletin,* 171: 1- 130.

Mosse, B. 1973. Advances in the study of vesicular – *arbuscular mycorrhizae. Annual Review of Plant Pathology*, 11: 171-196.

Parihar, D.R. 1993. Insect fauna of Khejri, *Prosopis cineraria* Linn. (Druce) of arid zone of India. *Ann. of the Arid Zone,* 33: 313 - 317.

Parihar, D.R and Singh, M.P. 1998. Insects associated with *Prosopis cineraria* in arid western Rajasthan, India. In *Prosopis species in the arid and semi arid zones of India*, edited by J.C. Tewari, N.M. Pasiecznik, L.N. Harsh and P.J.C. Harris. Published by *Prosopis* society of India and the Henry Double day Research Association: 99-102 pp.

Puri, G.S; Jain, S.K.; Mukherjee, S.K.; Sarup Shanti and Kotwal, N.N. 1964. Flora of Rajasthan. *Rec. Bot. Survey of India,* 19: 1 - 159.

Satyavir. 1996. Bruchid infestation of leguminous trees in the Thar Desert. *Tropical Science,* 36 (1): 11 - 13.

Sharma, Meeta and Ahmed, S.I. 2003. Relative toxicity of different insecticides against marwar teak defoliator, *Patialus tecomella* Pajni, Kumar and Rose (Coleoptera: Curculionidae). *Ann. For,* 11 (1): 127 - 132.

Sharma, Meeta and Ahmed, S.I. 2000. Biology and field efficacy of *Billeae atkinsoni* (Diptera: Tachinidae) a potential pupal parasite of marwar teak defoliator in arid and semi arid areas. *Indian Forester,* 4: 409 - 418.

Sharma, S.K. and Gupta, R.K. 1981. Effect of seasonal lopping on the top feed (loong) production and growth of *Prosopis cineraria* (Linn.) Druce. *Ind. J. For.,* 4: 253 - 255.

Verma, S.K and Satyavir 1995. Field insect pests of Rohida (*Tecomella undulata*) in Arid Zones of Rajasthan. *Ann. Arid Zone,* 34: 54 - 55.

WATER RESOURCE MANAGEMENT IN MARWAR REGION OF RAJASTHAN

Mahendra Mehta and Suneet Sethi
Jal Bhagirathi Foundation, Jodhpur – 342 003

ABSTRACT

In Marwar region, the daily temperature varies from a minimum of 4°C in winter to a maximum of 50°C in summer. The annual rainfall varies from 158 mm in Jaisalmer to 606 mm in Sirohi. The region has one deficit rainfall year in every three years. Evaporation from free water bodies is also very high i.e., upto 3.5 m/yr. However, due to excessive development of ground water in some pockets, decline of ground water levels has taken place. Repeated droughts and depleting sources of drinking water in the region are leading to mass exodus from the villages. Absence of sustainable means of livelihood has resulted in unemployment and increase in crime rate. The situation calls for a paradigm shift from drought relief to drought preparedness including risk management, public education, conservation, treatment and harvesting of available water resources. Traditional water harvesting structures like tankas, talabs, nadis, bawri and beri are the only solutions to fulfil the need of Marwar.

Introduction

The Marwar region of the Thar Desert in Rajasthan comprises seven districts *viz.*, Jodhpur, Jaisalmer, Barmer, Nagaur, Pali, Jalor and Sirohi. It is the most densely populated arid region in the world. The total population of the Marwar region is 12.2 million inhabiting a geographical area of 1,35,519 sq. km. (Census of India, 2001). The population density varies from 84-90 persons per sq km as against 3-4 persons / sq km in other deserts of the world. Apart from this, the cattle population is also very high with a density of 80 cattle per sq km. Thus, there is enormous pressure on the natural resources such as water and fodder.

Climatic features

The climate of the state varies from arid to sub-humid. To the west of the Aravalli range, the climate is characterised by low rainfall with erratic distribution, extremes of diurnal and annual temperatures, low humidity and high wind velocity. The entire region is characterised by hyperthermic conditions.

The annual rainfall in the state varies significantly. There is a very rapid and marked decrease in rainfall, west of the Aravalli range, making western Rajasthan the most arid part. The average annual rainfall in this part ranges from less than 10 cm in the north-west part of Jaisalmer (lowest in the state), 20 to 30 cm in Ganganagar, Bikaner and Barmer regions, 30-40 cm in Nagaur, Jodhpur, Churu and Jalor and more than 40 cm in Jhunjhunu and Pali regions and along the western fringes of the Aravalli range. The yearly total rainfall is highly variable with frequent spells of drought, punctuated occasionally by heavy downpour in some years.

In Marwar region, the daily temperature varies from a minimum of 4°C in winter to a maximum of 50°C in summer. The annual rainfall varies from 158 mm in Jaisalmer to 606 mm in Sirohi. The region has one deficit rainfall year in every three years. The irrigated area to the total geographical area varies between 0.62 per cent in Jaisalmer and 31.29 per cent in Jalore districts. The percentage of area irrigated by tube wells to the total irrigated area is above 85 per cent except in Jaisalmer where it is only 8 per cent. Evaporation from free water bodies is also very high i.e., upto 3.5 m/yr. The utilizable ground-water resources of the area comprising seven districts is about 2228 MCM/yr (Table 1). However, due to excessive development of ground water in some pockets, decline of ground water levels has taken place.

Table 1. Main features of districts under the Marwar region

	Barmer	Jaisalmer	Jodhpur	Nagaur	Pali	Jalor	Sirohi
Average annual rainfall in mm (1971-2000)	285	183	337	423	510	379	805
Average annual temperature in °C (Max. & Min.)	33.9 20.3	34.0 19.0	33.6 19.8	32.7 18.6	33.6 19.8	33.6 19.8	32.7 18.4
Major drainage system	Luni river	Ephemeral streams	Luni & tributaries	Luni river	Luni & tributaries	Jawai & Khari rivers	Luni & tributaries
No. of blocks	8	3	9	11	10	7	5
Ground water development in blocks (2004)	2-critical, 5 over-exploited, 1 safe	2 over-exploited	2 critical, 5 over-exploited, 1 semi exploited	2 critical, 7 over-exploited, 1 semi critical	5 critical, 5 over-exploited	7 over-exploited	3 critical, 2 over-exploited
Utilisable ground water resources in MCM/yr	256.46	60.09	375.64	548.37	282.16	432.33	272.74

Source: District Ground Water User Maps, Rajasthan, 2003. GOI, Ministry of Water Resources, Central Ground Water Board, Western Region, Jaipur.

Further the ground water quality is not good. The main features of ground water quality are given in Table 2.

Table 2. Ground water quality in Marwar region

Parameter	Number of villages/ habitations affected						
	Barmer	Jaisalmer	Jodhpur	Nagaur	Pali	Jalore	Sirohi
Chloride >1000 mg/l	743	251	336	463	185	151	28
TDS >2000mg/l	1070	383	605	811	313	281	95
Fluoride >1.5 mg/l	1170	467	799	1539	728	499	253
Total no. of villages/ habitations covered	1651	2172	1924	3375	1326	969	474

Source: Ground Water Quality Status of Rajasthan, 2001. P.H.E.D., Jaipur

Repeated droughts and depleting sources of drinking water in the region are leading to mass exodus from the villages. Absence of sustainable means of livelihood has resulted in unemployment and increase in crime rate. The State Government has spent crores of rupees on drought relief, mainly for generating immediate employment without addressing drought risk reduction. The situation calls for a paradigm shift from drought relief to drought preparedness including risk management, public education, conservation, treatment and harvesting of available water resources.

Natural Resources - Management and Traditional Wisdom

Genetic Resources

In spite of being a desert, the region has a diverse plant life comprising about 700 plant species including 107 grass species. The plant wealth is uniquely adapted to the hostile environmental conditions of the region. The sandy soils are low in organic content. Ground water is deep and often saline. Low rainfall, low water holding capacity of soil, high evapo-transpiration – all necessitate cultivation of drought tolerant crops that are present in plenty in the local germplasm. But tapping of ground water has enabled cultivation of water intensive crops in localized pockets that are harming the desert's ecology.

The Marwar region is home to the best breeds of cows, camels and horses in the country. The Marwari horse is known for it's long legs and ability to travel long distances. However, the livestock population is believed to be more than the current carrying capacity of the grazing lands. The requirements of water and fodder are so high that even during a good rainfall year there is little surplus. The current degraded state of the natural resources has arisen from improper land use planning and development strategies. Excessive drawing of ground water for intensive agriculture is leading to further water scarcity.

Eco-banks

The natural resources of the region are inextricably linked with the culture and form the backbone of the economy. These natural repositories are eco-banks, essential for the survival of life under the hostile conditions. Taking survival support from these 'eco-banks' for centuries, the society has shaped its way of life including livestock rearing, agricultural practices, food habits etc that culminated into region's culture, traditions, sociability and its ethos valued as their heritage.

Gauchars are village lands earmarked for grazing. Gravelly lands (*magar*) and hills (*bhakars*) are also common property resources open to grazing. Culturable wastelands (*parat bhoomi*) are owned by the government but the livestock can graze freely. Agors are protected catchment areas of water bodies that are also open to grazing. *Birs* are forest lands owned by the government. Forest products like fuel wood, fruits, leaves, herbs etc. are collected from the forests and are important for the survival of poor communities. Besides these natural resources, most villages have sacred groves called *orans* that are dedicated to local deities, heroes or martyrs.

The agor-gauchar-oran (AGO) matrix has been developed, protected and controlled by peoples' institutions for centuries. AGO development is essential for supporting rural poor on a sustainable basis to lessen the impact of droughts and famines in the desert. Together these resources provide water, fodder, firewood, fencing and thatching material, herbs, resin, gum, fibre, timber, non-edible oils, natural dyes, tannins and raw materials for crafts which together constitute the lifeline of the village economy. In the past, these natural resources were conserved by unwritten code of conduct and even in the present state of erosion of traditions, these continue to exist in a depleted state.

The presence of *agor-gauchar-oran* trio in desert villages is a unique example of surviving support system in an inhospitable and stressed ecosystem. These community resources are the main source of free fodder. Forests in the desert are bare minimum not exceeding 1.8% of land, while AGOs together constitute about 5.6% of the total land. The latter is thus a very important resource but needs upkeep, maintenance and improvement to enhance productivity to meet the growing demand for fodder.

In the region having less than 250 mm of annual rainfall, animal husbandry is the main occupation, the livestock sustains on common grazing land and fallow fields. There are sections of the community that have large herds but no lands and their livestock are entirely dependent on the common pool resources called '*gauchars*'. In the area receiving more than 250 mm mean rainfall, mixed agriculture is practiced. Most farming households maintain a sizeable livestock, which is sustained on crop residues, on fallow and common grazing lands. Double cropping is practiced where ground water is available.

Traditional Water Management

The desert society has evolved indigenous technologies of water management for sustaining the human and cattle population. The techniques include *talab*, *nadi*, *bawdi*, *khadin, tanka, beri* and rooftop rainwater harvesting systems. But wherever piped water supply becomes available, the traditional water harvesting systems are abandoned.

In the past, Marwar had a great tradition of community management of water resources. *Talabs* were constructed by the participation of the whole community. The community contributed in cash, kind and labour. Since water could not be transported from far off places, they took good care of their local resource. Rules were formed for proper utilization and management of the resource. Besides *talabs,* there were ingenious water harvesting technologies used, like *bunds* and underground storages catering to palaces or even whole habitations. At a smaller level, each household would collect rainwater in *tankas* or *kunds*.

Traditional water harvesting system have been severely eroded, thrown into disuse and even eliminated. The existing water bodies have silted mainly due to deforestation, degradation of catchment areas and dependence on ground water has increased manifold. At times, encroachment of tank bed has taken place, due to which even the ground water recharge has reduced considerably. Some of the reasons for decline of traditional systems are:

- Increase in population resulting to greater demand, which cannot be met with traditional systems.
- Availability of modern and convenient alternate water supply.
- Greater dependence on ground water, which is under the control of individuals.
- Subsidy on for power for withdrawal of ground water.
- Lack of motivation, community ownership and management aspects in Government schemes.
- Commercialisation of agriculture and cattle rearing.
- Lack of ownership of common resources.
- Lack of support by Government for traditional water bodies.

- Politicising the community for common resource management.
- Caste and gender biased society affecting both access and management by marginalized sections.

Methodology and Strategies

Initiating local action for reducing vulnerability to drought, the Jal Bhagirathi Foundation (JBF) has already mobilized communities in the Marwar region to revive traditional water management practices. This program provides immediate relief to a region suffering from recurring drought and mobilizes communities to build their capacity to undertake long-term drought proofing of the region.

The main problems identified in development and sustainability of water harvesting systems in this region are:

- Low rainfall – 250-350 mm/yr.
- Less number of rainy days – 10 -15 days.
- Frequent drought.
- High evapo-transpiration – 2500-3000 mm/yr.
- Highest density of human and livestock population in desertic areas of the world.
- Ground water quality is not potable due to high salinity and at places due to high fluoride.
- Negligible to NIL water availability during peak demand season (April-June) - transportation of water through tankers and requirement for *tankas.*
- Flat topography – less runoff as sheet flow, not as defined channels – harvesting is complicated.
- Presence of salt pan at shallow depth makes *tanka* construction more difficult.
- Community is very poor.

In this region, *tankas* have also been constructed by Panchayats/Government, but these *tankas* have no provision for harvesting rain water. They are only meant for storing transported water. At present, many of these *tankas* are damaged/ leaking and are out of use.

JBF is constructing *tankas* in schools by incorporating roof water harvesting. Special attention is given to condition of roof, design of conveyance system, installation of good quality hand pumps for hygiene (as against drawing water by buckets), provision of silt chamber and overflow in *tankas*.

JBF also mobilises the people to undertake construction of community *tankas* on common lands. These *tankas* are also constructed for harvesting rainwater. Though both school and community *tankas* are designed for rainwater harvesting, the rainwater helps them to tide over a few months, after which water is transported from other sources and stored in these structures. If water supply is available in nearby sources such as *nadis*, people use the nadi water as long as it lasts, and then they use the rainwater harvested in the *tankas*.

Results and Discussion

During the peak summer season the community resorts to transportation of water from village *talabs* or private tube-wells. The Government also supplies water to the water scarce habitations during the summer months. The JBF initiated a water supply programme during the summer of 2005. Economic analysis of the two-month long water distribution programme is given in Table 3.

Table 3. Drinking water supply - economic analysis of water sulpply for 2 months

S.N.	Description	Agolai	Balotra	Luni	Rohet
1	Date of commencement	5.5.05	10.5.05	10.5.05	6.5.05
2	Total No. of days of water supply (days)	51	47	52	48
3	Total volume of water supplied (L)	15,65,000	9,17,500	19,68,000	44,05,000

4	Total no. of households benefited (HH)	513	440	1,680	1,320
5	Total distance covered (Km)	11,341	9,627	6,044	16,368
6	Total cost	1,62,150	1,41,285	2,17,937	2,14,410
7	Volume of water supplied per household (L)	3,051	2,085	1,171	3,337
8	Volume of water supplied per household per day (L)	59.82	44.37	22.53	69.52
9	Cost of water supply per household (Rs)	316.08	321.10	129.72	162.43
10	Cost per household per day (Rs.)	6.20	6.83	2.49	3.38
11	Cost of water sup-ply per litre (Rs.)	0.10	0.15	0.11	0.05

The cost of transporting water varied between five to fifteen paisa per litre. The cost of water transported is variable because of following factors

Distance from source

Type of road – more for kuccha roads

Capacity of tanker – controlled by road condition

Availability of water as free or on payment.

Scientific and Technological Improvements: Initiative of JBF

The Foundation has initiated following technical improvements in its programme

- Training of staff for scientific design of rainwater harvesting structures.
- Scientific design of *agor* based on rainfall and *tanka* capacity.
- Computation of run-off.
- Standardisation of rates of desilting/ excavation of *nadis/ talab.*
- Taking test pit while deepening of talab.

- Motivating community for construction of deeper surface water bodies to minimise evapotransp-ortation.
- Design of pal and beam.
- Standardisation of *tanka* dimensions based on gross capacity.
- Provision of silt chamber and air vent in *tankas*
- Design of feeder channel.
- Design of conveyance system for roof rainwater harvesting.
- Standardisation of bill of quantities and rates of various items for block and clusters for quick preparation of estimates.
- Development of system for maintenance of micro-projects.
- Development of system for post implementation evaluation of micro-projects.
- Motivation of community for desalination of ground water.
- Motivating community for construction of RCC/PCC *tankas* which are 30% cheaper than stone masonry *tankas* and are more effective where there is problem of swelling clays as shown in Table 4.

Table 4. Economic analysis of stone masonry and RCC tankas

Type of Tanka	Cap (m^3)	Av. Cost in (Rs.)	Life Yrs.	Av. No. of Benifi ci-aries	Av. Water req. (LPCD)	Av. Per. Day req. (Lit.)	No. Days it can serve Benifi-ciaries	Cost. saving @ Rs. 300 Per Tanker (4000 L)	Total saving in 20 Yrs. (Rs.)	Net saving (Rs.)	Av. Cost of Water per lit. in Paisa
A. Stone Masonry	20	35,000	20	50	10	500	40	1,500	30,000	(-) 5,000	9
	30	45,000	20	70	10	700	43	2,250	45,000	0	7

	40	55,000	20	80	10	800	50	3,000	60,000	5,000	7
	50	70,000	20	100	10	1,000	50	3,750	75,000	5,000	7
	60	85,000	20	100	10	1,000	60	4,500	90,000	5,000	7
B. RCC/ PCC	20	23,000	20	50	10	500	40	1,500	30,000	7,000	6
	30	30,000	20	70	10	700	43	2,250	45,000	15,000	5
	40	35,000	20	80	10	800	50	3,000	60,000	25,000	4
	50	45,000	20	100	10	1,000	50	3,750	75,000	30,000	4
	60	55,000	20	100	10	1,000	60	4,500	90,000	35,000	5

Some of the other standards like dimensions of *tankas* for various capacities and requirement of minimum catchment (agor) as worked out is given in tables 5 & 6.

Table 5. Standard design of tanka

	For Normal Area			Area where salt pan and/or hard formation at shallow depth	
S N	Capacity (KL)	Diameter Inner (M)	Depth Clear (M)	Diameter (M)	Depth (M)
1	60	3.70	5.50	4.20	4.35
2	55	3.70	5.15	4.50	4.00
3	50	3.70	4.65	4.20	3.60
4	45	3.70	4.20	4.20	3.25
5	40	3.70	3.70		
6	35	3.65	3.35		
7	30	3.65	2.90		
8	25	3.65	2.40		
9	20	3.65	1.90		

Table 6. Computation of agor area for community tankas

S.N.	Capacity (KL)	Min. size of Agor (Square M)	Radius Agor (M)	Remarks
1	20	118	6.00	Av. Rain Fall = 300 mm
2	25	147	7.00	Monsoon rain fall (80%) = 240 mm
3	30	176	7.50	Runoff coeff. = 0.70
4	35	206	8.10	
5	40	235	8.70	
6	50	294	9.70	
7	60	353	10.60	

Case Studies

Bhandiawas, Karnisar Talab

Village Bhandiawas lies 6 km south-west of Pachpadra town. This is a medium-sized village famous for its temple of Karni Mata. There is a *talab* in this village known as Karnisar after the local deity Karni Mata.

In comparison to its catchment area, the size of the *talab* was too small and when there was good rainfall, there was always a danger of the village getting flooded. During the summer of 2004, JBF had started working in this area when a local leader Shri Naindan Charan approached JBF and informed them about their problem.

The Karnisar Talab was surveyed and the work requirement was estimated for Rs.4,00,000. The Jal Sabha collected 25% of the estimated cost as community contribution. Within 20 days, the small water body was converted into a large *talab*. The villagers were so happy with this work that they invited Maharaja Sahab and Maharaniji to their village.

Though the rainfall that year was average, a good quantity of water was collected in the *talab*. Additional storage of 17800 cu m was created to harness surplus runoff in the *talab*. The Jal Sabha decided to appoint one person to safeguard the *paal* and to check wastage of water. It was decided that if a person from any other village fills water from this *talab,* by tanker, he should pay a small amount of money. Revenue so raised could be used to pay for a watchman to take care of the *talab*. Any person wanting to fill his tanker from this *talab* gets a token from this village for Rs.20. In this way, the Jal Sabha has collected Rs. 30,000. With this money, the Jal Sabha has decided to do more work in the *talab*.

Community empowerment - Kheechion ka dahipada, Luni Block

In this village, the community has constructed the Jorawar Talab with JBF's help. The unique feature of this talab is the beautiful system of inlet channels constructed in the catchment area. The Jal Sabha has developed an exemplary system of maintenance for this talab and recorded the rules in its register. The rules are – women of the village clean the feeder channels during *amavasya* and men do the cleaning on *ekadashi*. Rules are also formulated to regulate water collection from the talab. Water collection by headloads or camel/bullock carts is permitted, but if anybody fills water in tractor tankers, that person has to pay a fine of Rs.1100. Two watchmen have been employed by the Jal Sabha to see that water is not collected by tankers.

In the month of February, a group of about 15 people came from the neighbouring Khejariali village and forcefully tried to collect water in their tankers. A meeting of the Jal Sabha was called and it was decided to report the matter to the police. The police did not take action. The same incident happened again, but this time the villagers engaged the intruding outsiders in small talk, and few other people went to the police. The police saw for themselves that these people were collecting water against the rules of the Jal Sabha.

The Tehsildar, Patwari and Superintendent of Police were called and they decided that since only the inhabitants of Kheechio ka Dahipada had contributed for development of the talab, they had the authority to form rules for its utilization. The villagers filed a case against the intruders. Meanwhile there has been no further incident of breaking rules in this village.

Gender Concerns in Water Management

Although the role of women in management of natural resources in general and water in particular, is accepted but involvement of women as equal participants in water resource management projects has been achieved in a very few cases. All significant decisions - the location of the water source, the rules for its use, maintenance of water quality, the application of penalties for misuse, the distribution of water among competing individuals or interests - are taken by men or male-dominated bodies.

The impact of drought is hardest on women because of their socio-cultural and economic positioning within the family and community. When men lose their jobs, they migrate often leaving their families behind and women have to take up the additional task of earning for the family. In the job market, women are constrained by lack of education and skills, assumed lower working capacity, lesser social sophistication and are at risk of sexual harassment. In Rajasthan, it is a common site that more women work on famine relief jobs than men, the reason- how can men work for such low wages.

It follows that women have the greatest stake in water availability and therefore, it is essential to include them in the decision-making process. In India, it is estimated that the national cost of women fetching water is 150 million women weekdays per year, equivalent to national loss of income of 10 billion rupees per year.

Summing Up

In the Marwar Region, availability of water is limited in time and space. In order to provide sustainability to drinking water supply, following measures are to be taken on priority

- Motivation of community for rainwater harvesting.
- Identification and revival of big talab as drinking water security for 5-6 villages.
- Cleaning of Agor (Catchment) and construction of feeder channels for augmenting water inflow in talabs.
- Motivation and construction of private schools & community tankas with Rooftop Rainwater harvesting and/or artificial Agor for water security.
- De-silting/ renovation of existing water bodies.
- De-salination/ de-flouridation of ground water.
- Motivation of community for upkeep of traditional water harvesting structures.
- Better involvement of women for planning, maintenance and use of water.
- Recharge of Ground Water especially through existing wells low cost technology.
- Motivation of farmer for less water intensive crops & adopting modern methods of irrigation eg. Sprinklers, Drip irrigation.
- Creating awareness among women and children about water conservation and water management.

References

District Ground Water User Maps, Rajasthan, 2003. GOI, Ministry of Water Resources, Central Ground Water Board, Western Region, Jaipur.

Ground Water Quality Status of Rajasthan, 2001. P.H.E.D., Jaipur.

FOREST TECHNOLOGY EXTENSION - INTEGRATED APPROACH

G.D. Singh, S.K. Sharma and J.P. Lakhera
Directorate of Extension Education,
Rajasthan Agricultural University, Bikaner (Raj.)

ABSTRACT

Extension of promising Technologies in agriculture and forestry play major role in bringing about positive changes in agriculture and allied production. However, long time Technology Extension in agro-forestry and forest management could not gain momentum on account of one or the other reason. Obviously, looking to the problem of afforestation and forest products in fragile arid agro-eco system and more specially degradation of environment, forest research extension is becoming precious. Considering degradation of natural resources by exploitative agriculture and environmental degradations, drastic refinement in the research of farming systems in agriculture *viz.* forestry, as well as, technology dissemination system are needed. Further, real food security may warrant effective coordination between production systems and value addition activities. Efficient extension system in rural areas may create wonderful avenues of employment through varied forest products – fuel, fodder, medicine, gums, resin, fiber, rough paper, kathha, lac, furniture timber, plates and bowl, packing boxes, paper bags, khas mats, broom, honey and wooden toys.

Therefore, to secure healthy integrated forestry technology extension certain scientific measures will have to be incorporated in dissemination picture. Now the time is mature to empower farmer and farm women to act as agents of information on one hand and as a system partner on the other in a broad base fashion. The transferable forest technologies will have to be weighed on the PHT and value addition balance and if this is done, I am sure the client will rush for the technology. Precisely, the new forest extension policies for integrated forest technologies and information should consist of: (i) environmental research and extension linkage; (ii) capacity and skill; (iii) gender empowerment; (iv) information technology; (v) market extension and value addition and (vi) management.

Introduction

Our forefathers called forest as "Bhomiji" (*Van devta*), worshiped them at festivals and relevant occas-ions. About 200 million country population are wholly dependant on forest resources for their subsistence. Social activists like Sh. Sunder Lal Bahuguna, Ms. Medha Patkar, as well as, public movements like "Chipko" had been very important source of extension in forestry.

Extension of promising technologies in agriculture and forestry play major role in bringing about positive changes in agriculture and allied production. However, since long time technology extension in agro-forestry and forest management could not gain momentum on account of one or the other reason. Obviously, looking to the problem of afforestation and forest products in fragile arid agro-eco system and more specially degradation of environment, forest research extension is becoming precious. Considering degradation of natural resources by exploitative agriculture and environmental degradations, drastic refinement in the research of farming systems in agriculture *viz.* forestry, as well as, technology dissemination system are needed. Further, real food security may warrant effective coordin-ation between production systems and value addition activities. Efficient extension system in rural areas may create wonderful avenues of employment through varied forest products – fuel, fodder, medicine, gums, resin, fiber, rough paper, kathha, lac, timber of furniture, plates and bowl, packing boxes, paper bags, khas mats, broom, honey and wooden toys. Govt. of India has paid little attention to forest extension programmes as compared to crop extension. As per ICAR report of NATP (1999) only 12 I.V.L.P projects were sanctioned in forestry against 923 projects in crop production, at all India level from World Bank assistance.

Probably on account of deficit forest extension measures during green revolution period in canal area, large number of trees existing in crop fields were removed as a safe guard against bird damage to food crops. Later on, due to widespread problem of insect – pests in crops, specially, cotton and degradation of natural resources including soils, farmers and scientists realized the importance of trees in crop fields. Obviously, policy makers in agriculture have started floating the idea of evergreen revolution necessitating presence of trees and shrubs all around and fields.

Past Technology Extension Programme

A large number of community development technology extension programmes were floated since 1950.

CDP (1952)
NES (1953)
IADP (1961)
IAAP (1964)
HYVP (1966)
ETEP (1966)
SMED (1969)
ND (1970)
T & V (1970)
NWDPRA (1990)

A number of these programmes remained failure or produced limited success on account of limited approach and objectives. They were mostly silent in holistic and integrated approach on environment conservation and transfer of technology on farming systems. Moreover, programmes on HYVP, N.D. and T & V system however, recorded good result in green revolution in the irrigated areas, but the environmental protection aspect and afforestation were left untouched. Past programmes generally carried out by public extension had major limitation of top down approach and non-consideration of environmental relations including horticulture and forestry. The farmer was given less weightage and as such a feed back based on clientele demand driven and accountability never developed.

Also technology extension in natural resource management spheres say, sustainable farming system, forestry, LPM and fisheries were not attended seriously to and the programme failed.

Integrated Forest Technology Extension

Extension is a set of function – forest training, I.T. on forestry, forest technical skill, organization, credit, incentives, forest planning, marketing, monitoring and evaluation. Therefore, to secure healthy integrated forestry technology extension, certain scientific measures will have to be incorporated in dissemination picture.

(1) Broad base programming - Farming system approach (A.F)

(2) Farmer participation (PRA)

(3) Post harvest technology and value addition (Rural base small scale industries)

(4) Technology application to commercialization (Quality, marketing and export)

Now the time is mature to empower farmer and farm women to act as agents of information on one hand and as a system partner on the other in a broad base fashion. The farmer is very intelligent and he will like to judge economic aspects of the technology. Obviously, all the transferable forest technologies will have to be weighed on the PHT and value addition balance and if this is done we are sure the clientele will rush for the technology. Precisely, the new forest extension policies for integrated forest technologies and information should consist of:

1. Environmental research and extension linkage
2. Capacity and skill
3. Gender empowerment
4. Information technology
5. Market extension and value addition
6. Management

Discussion

I. Environmental Research and Extension Linkage

Utilization of waste land for forest development and environment conservation may prove a wonderful aspect of technology in forest management worth speedy dissemination. Large number forest trees say from Shami in arid lands to Poplar in semi-arid to sub-humid areas are drying for one or other reason. Diversified research attempts are needed to protect forest bio-diversity to sustain fragile environment. It has been established that a dense forest land invite more rainfall leading to sustain bio-diversity. CSSRI is very successful in raising trees in salt affected soils through bore hole gypsum technique.

The most widely talked research and extension linkages should be carried out in realistic approach rather than simply as a traditional formality. Precisely a healthy research extension and formal linkage must be established at priority to refine bottom up feed back in diversified forest problem areas so that location specific technology recommendations are developed. Obviously, such demand driven scientific recommendation will definitely solve the problems of the farmers related to forest production, farming system, agro-forestry and silvi-pasture.

For example, farmers of arid eco system generally indicated a feed back on low productivity of pods (sangari) from *Prosopis cineraria* trees on account of defective lopping schedule. It is generally observed that farmers are lopping their Shami trees for fodder "loong" in October-November and as such, these trees would not bear sangari in summers on new growth. Scientists of farming systems conducting research on this feed back have come forward with useful forest technology on alternative lopping schedule producing high yield of sangari. Skipping lopping of trees in October – November can harvest good quality of fodder (33 kg) and fruits (15 kg) in summers.

Based on another feed back, Scientists have also improved B.T. of Shami trees with thornless trees innovating wonderful breakthrough in vegetative propagation. Such beneficial technology and that of benefits symbolizing 5 F (Fodder, Fruit, Fuel, Food & Fertilizer) principle is needed to be transmitted through research extension linkage.

Table 1. Viable technological intervention in Forestry based on demand driven approach

Forester demand	Technology Intervention	Source
High Pod yield with fodder (*loong*) from *Prosopis cineraria*	Changing lopping schedule of Shami trees from October to April-May has been found promising for high yield of both *loong* and pods (Sangari).	I.C.A.R News (2003)
Thornless trees of Shami	Technology of Vegetative Propagation developed for thornless trees	R.A.U., P.C (2005)

Obviously, EAC, ZREAC, Kisan Mela and FMD are to be utilized more efficiently to get break through in Research Extension – Farmer linkage and fruitful results. Advance research and extension are needed to build capacity of farmers and farm women, such as:

1. Identify special product group farmers,

2. Environment consciousness – Training, practical etc. print media,
3. Training in low cost input – Training, special product cultivation modules, PHT strategies say harvest, treatment chemicals,
4. Creating PHT facilities in the vicinity of product sites,
5. Training on PHT processing and value addition (storage safety) and
6. Use of films, T.V and internet for creating awareness about product details.

II. Capacity and skill

As far as effective dissemination of advance technology is concerned, we have to design specialists with diversified skill and capacity to be ready to work at village level. We may start farmer schools at district level to achieve such objectives. A subject matter specialist needs to be refined with such knowledge and skill that he can solve the problem of a village forest. Accordingly a team of specialists / experts with technical skill and knowledge of information net work will serve a ring of technology dissemination chain between extension officer and local farmers.

All extension functionaries needs thorough training on long term basis consisting of environment, foundation courses, need assessment, agro-forestry, PRA, agri-business, WTO, marketing, PHT, communication, data handling etc. Probably this is the reason that OFT (On Farm Testing) should be conducted very efficiently by involvement of farmer and scientists at KVK.

To manage training infrastructure in an efficient manner, all training centers at state level should be net worked with H.Q. SAU and AAFRI. All ZTC attached to research stations and KVKs may be utilized efficiently to train state level extension functionaries. Agricultural Universities may depute training associates, time to time, to

National Centers of training like AAFRI manage etc. for the sake of capacity building. Large numbers of Agro-clinic-cum agri-business centers should be opened for the service of farmers.

III. Farmer empowerment

Farmers should be involved in planning and implementation of extension programmes through different institutions. Most of the field programmes as watershed development, forestry, horticulture, value addition and product marketing may be implemented through farmer user groups as JFM or watershed committee and so on. Formation of SHG and TTC is the important step of NABARD in the line of farmer's empowerment. Therefore, a new look must be given to training, demonstrations and information system.

ICAR has started giving special attention to farm women empowerment with reference to LPM, food processing, value addition and house hold occupation including marketing. Certain functions are to be redesigned to focus farm women – say; sensitization of extension officers towards female power, special training for women, increase involvement of farm women as researchers, extension officer and marketing for appropriate technology to boost female attendance in forestry and allied programmes. Ensure adequate representation of female in training programmes specially designed for occupations suiting women say – collection of fuel, gum, kathha, lac, honey, preparation of plates or bowls, boxes, paper bags, mats, broom, wooden toys and creating special drive of training of lady extension officers.

Stakeholders must be empowered by vocational trainings such as bee keeping, forest and watershed management, plant propagation and nursery. Initiation of financial help from NABARD is also a welcome step of farmer empowerment. Nearby agri-business centers should be well equipped with nursery and forestry inputs to be available on sale. NABARD may come forward to help SHG in forestry enterprise including processing and value addition of forest related products.

IV. Information technology

Fundamental notes regarding forest needs, environment and care may be included in school and college syllabus. The science of information technology is opening new vistas in agriculture extension by bringing the farmers more close to agriculture scientists, AAFRI, extension officers, the ATIC and ARS as well as KVK. Special steps must be taken to harness the fruits of I.T for farmer's empowerment and technology transfer:

1. I.T. Network – extensive use
2. Agricultural marketing and I.T. – advance markets
3. Use of EMM in extension – say T.V./Radio/THL
4. I.T. support – NET ARIS, Web site – ATIC
5. Kiosks – Private P.C.O., e-mail (KVK)

On the lines of advance countries, Government may improve information flow and quality of information to farmers by training and publications to facilitate transaction and also sponsor workshops for farmers with agri-business enterprise to improve mutual understanding of market.

After establishing the service of telephone helpline, KVKs and institutes should be equipped with e-mail, internet and web site connections for the benefit of farmers. Accordingly, use of I.T. facilities must be encouraged in training, FMD, reporting, meetings, kisan mela, workshop etc.

V. Market Extension and value addition

A lust for good employment and higher income which is warranting the process of diversification in agriculture, call for greater generation and transfer of good technology. With the air of liberalization and broken trade barriers in post WTO regime, we have a major shift from production led agriculture to market driven agriculture, giving more stress on quality aspects.

Market extension services are important to assist farmers in marketing of their product in the congenial trade environment. Obviously, market extension message may include:

1. Advice on product planning
2. Market information
3. Secure markets
4. Alternate markets
5. Improved PHT practices – grading, packaging etc.,
6. Cooperative market approach.
   - For securing export market quality aspects of product are very important.
   - For efficient market extension coordination with MANAGE, NABARD, NIAM, IRMA, IIFT and APEDA could be established for sharing experts in capacity building of extension institutions.

There are a number of forest products which are vital for international trade. Therefore, to project our product in export market certain extension strategies are much pertinent to educate the farmers say- SWOT analysis.

1. Strength – Diversified climate, special products, labour and technical man power.
2. Weakness – quality, infrastructure (strategies), cost of transport, PHT losses, lack of PHT know how.
3. Opportunities – GATT, WTO, Organic Products, more practical in arid and semi-arid areas.
4. Threats – Stiff competition from internal markets and external markets of developing countries like Brazil, Bangladesh and Sri Lanka.

To divert the attention of farmers towards a system, the system must be fool proof as far as its documentation or demonstration is concerned. For example a profitable agroforestry or forestry system must be taken to the farmer as Front Line Forest Demonstration or as a farming system model. In arid and semi-arid areas any body may see a number of economic agro-forestry and agro-horticulture models. In arid areas an important plus point with the final (value addition) products is purity in nature.

To add more attraction into a forestry system, the end products may be refined in quality and further transformed in final valuable products through processing functions. Obviously the value addition treatment will further increase the profitability of a farming system.

1. Agro-forestry system of Shami + Guar

   Value addition products – Sangari (pods for vegetable), gum, resin, *loong* (fodder), feed block.

   Acacia catechu + Mustard

   Products – Katha, oil, oil cake, honey, toys.

2. Forestry system of pipal, Tendu leaf

   Product – feed block, plates and bowl.

3. General trees -

   Product – Paper, packing boxes, paper bags, mats and toys.

4. Mixed stand consisting of palm, vetiver, saccahrum sp.

   Product – Vetiver mats for cooler, brooms, scent, chord, mudha and fans.

5. Agri-horticulture – Tomato, guar, cucurbits, alloevera in ber and anola.

   Product – Dry vegetable, jam, jelly, chutney, pickle, powder and others.

There may be several other system and resultant value added food products depending on the agro-climatic reservations of the region. For marketing information close links and linkage should be established with marketing hubs engaged in similar products including district level Industry Officers, NGO and KVIL centers.

VI. Management

Revised forest policy 1986 warrant priority to conservation and livelihood requirements of forest stake holders. Obviously, JFM emerged an important tool and that also extension tool with an objective of protection of forest and environment through joint efforts.

In some areas, JFM are doing excellent such as in World Bank Projects, wages of labour for plantation of saplings are major attraction of people participation.

Management of extension and dissemination of relevant agricultural technology including farm forestry for wider application and increased productivity, would necessitate a holistic integration of ICAR, National Centers, public, private organizations, NGO, SAU and advance centers of agriculture and ultimately the farmer on the lines of JFM. Such aspects are:

1. Support of center Govt. to State Extension Service carefully managed.
2. Govt. assistance to institutions for farmer's programmes be managed and monitored properly.
3. Panchayati Raj Institution may be utilized for training environment and forestry mandates if managed judiciously and timely.
4. CGAEF – funds from ICAR to address specific research and extension needs of farming system of a watershed / village / block be managed properly.
5. Revolving fund – for plant material, seed, seedlings and allied inputs for extension institution.
6. If certain extension center is not giving desired results, institutional restructuring may be done.

Conclusion

Center for Environment Education under NFFD is doing commendable job towards environment know how at grass root level. Large numbers of such institutions are the need of this hour to disseminate forest and environment conservation technology. Foresters, farmers and all stake-holders attached with forestry and environment protection should be properly trained for dangers of environment degradation including dangerous natural calamities including adverse climate. In this era of intensive agriculture large and well to do farmers are no more dependent on public extension system, however, small and marginal farmers and land less labours will continue to depend on Government supported extension system. Recently, in the light of WTO, agriculture demands a system of market led extension with impetus on crop diversification including agro – forestry, allied farming systems, post harvest technologies, value addition and export agriculture. In fact, for more efficient dissemination of advance research, the present day research and extension feed back should be driven by farmers on location specific priorities. Obviously FLD, trainings, field days, FMD, Kisan Mela, T.V., Radio, telephone and meetings will continue to be important agencies of technology transfer. Generally the farmers before contacting with an innovation, will seek the pros and cons of profit margin with the innovation.

Literature consulted

Anonymous. 1999. Project report on NATP, ICAR, New Delhi.

Gill, A.S. 2003. Role of Agri-Horticulture in Agroforestry-A Success Story, ICAR News, Vol. 9 (4): 8 & 9.

Pareek, O.P. 2003. Growth regulation in Khejri for annual production of both Sangari & Loong, ICAR News, Vol. 9 (4): 18.

Roy, G.L. 2001. Systems of Extension in India; *Extension Communication and management*; 170-217.

FORESTRY EXTENSION IN HOT ARID REGIONS OF INDIA: SOME PERTINENT ISSUES

Pratap Narain and J.C. Tewari
Central Arid Zone Research Institute, Jodhpur 342 003

ABSTRACT

With emergence of plough, man developed agriculture as a stable profession as a source of sustenance. However, forests/ trees have always remained inter-linked in complex web of man-agriculture-trees-livestock continuum. The hot arid regions, which are spread over 3.17 million ha (61.0% of which lie in arid western Rajasthan), farmers have for many generations maintained tree-crop combinations as a traditional stable land use. Historically, forestry and agriculture sectors have remained isolated. However, society needs both. The agriculture extension system has a sound base in the country and contrary to this, forestry extension strategies are still very weak. In fact, need of hour is to inter-link agriculture and forestry extension in such a way that both the sectors flourish simultaneously. Approaches like Basic Assessment Guide (BAG) and Participatory and Reflective Analytical Mapping (PRAM) are very effective for forestry extension projects, however, in farming system scenario, both agriculture and forestry sectors have their own importance. The paper attempts to discuss various tools and approaches possible for better forestry extension in hot arid regions and importance of inter-sectorial approach.

Introduction

Millions of people representing a great variety of cultures and landuse practices live in or on the edges of tropical forests (Isager *et al.*, 2002). Some groups practice crop rotation in permanent fields and other sustain themselves through various forms of agroforestry, some times in combination with other sources of income. Many are shifting cultivators, who have for generations lived in and used forest resources. In fact, with emergence of plough, man made agriculture as a stable profession and a basic source of assured supply. However, the forests/trees and their products have always remained interlinked with man-agriculture-trees-livestock continu-um in one way or the other. Forests or trees on crop field or trees outside the forests and crop fields are important component of energy fluxes in any agro-ecosystem (Tewari, *et al.* 2003).

In hot arid region of India, which are spread over in 31.7 million ha. farmers have maintained agroforestry as a traditional landuse system since time immemorial, which is characterized by deliberate retention of woody taxa on crop fields or fallow lands. Because forest cover in hot arid regions of India is very low, these trees/shrubs fulfill demands of forest products from trees (Khan *et al.* 2006). This paper attempts to discuss the key issues for extension forestry in hot arid regions of the country, especially of arid western Rajasthan, which covers 61% of hot arid regions of India and designated as principal hot arid region.

Pattern of Woody Vegetation and Issues

The forest cover in arid western Rajasthan is reported to be not more than 1%. Tree cover designated as forest cover is represented by *'Orans'* and *'Gauchars'* and no area has crown density >40%. According to Narain and Tewari (2005), the woody vegetation in upper half of arid western Rajasthan follow a rainfall gradient in which *Prosopis cineraria* – *Acacia nilotica* dominated >400 mm rainfall zone; *P. cineraria* 300-400 mm rainfall zone and *Ziziphus* spp. – *P. cineraria* 200-300 mm rainfall zone. In <200 mm rainfall zone *Ziziphus* spp. – *P. cineraria* – *Salvadora* spp. dominated. In lower half of arid western Rajasthan different set of woody vegetation dominated the scene in which *A. nilotica* var. *Cupressiformis, Azadirachta indica, Salvadora* spp. etc. were the major species of traditional agroforestry systems (Anonymous 2003).

Farmers allow to climax utility vegetation to grow on their crop fields for their varying needs. In addition to traditional agroforestry system related woody vegetation, man made utility conservation forestry in hot arid region include:

- Sand dune fixation
- Shelterbelts
- Field boundary plantation
- Plantation in IGNP command

Participation of local communities with respect to land conservation in said forestry related programmes is often described as associated with concept of community participation. However, in true sense, the participation of local communities in those activities is much below the desired limit. In fact, community forestry means that a forest or a common property resource (CPR) is managed by the people, who live close to those units, but scenario in arid western Rajasthan is altogether different. On papers, CPRs are managed by the local communities, however, such units have become no man's land and everybody exploit such important resources without giving due consideration for their restoration. In the process, thousands of CPRs which were once the fodder repository for livestock have been degraded to an alarming state. Trees, shrubs and herbaceous cover has been deteriorated to the extent, from where recovery of ecosystems is very difficult. The ever increasing livestock population and associated demands are major factor responsible to this state of degradation (Narain and Tewari, 2001).

Agriculture Extension vis-à-vis Forestry Extension

Achievement of self reliance in food grains is the greatest success of agricultural extension with solid support of research system (Shekhara, 2001). Agricultural extension base in the country is quite sound as compared to forestry extension services. Still, forestry is considered a landuse for residual lands. Such a mind-set for forestry is not going to pay any dividend in future. If the forestry is seen in farming system perspective, an altogether new picture emerges. India accounts for mere 2 per cent of world's total land area, 1 per cent of its forests and a miniscular 0.5 per cent of its pasture lands; yet the country sustains 18 per cent of world's human population and 15 per cent of its livestock (Pande, 1995). Estimates indicate that some 50-60 million Indians rely on forests or tree resources for their subsistence.

In hot arid tract of western Rajasthan more than ten million people are dependent on trees outside the forests. This includes tree on crop fields, woody vegetation on wastelands or so-called revenue lands, gauchars, canal bank plantations, etc. Thus, trees/shrubs are intricate part of life support system in arid ecosystem. Even then, there is no forestry extension system for hot arid regions. Trees on crop fields, which are major source of fuel (still more than 70% rural population is dependent on tree/shrub species for fuel energy) and fodder (man: livestock ratio ranged from 1.0: 0.75 in eastern margin to as high as 1.0: 4.0 in extreme western fringes) are never considered as forest wealth. The concentration of forest extension agencies is limited to canal bank plantation in IGNP command area. Therefore, it is the need of hour to develop a sound forestry extension base for hot arid tropics, which primarily focuses the strengthening of woody components of traditional agroforestry system besides the plantation strategies on wastelands, permanent fallow lands, pastures, road sides, canal banks, urban areas, etc.

Approaches for Forestry Extension in Hot Arid Regions

With respect to forest conservation, participation is often associated with the concept of community forestry, a forest managed or co-managed by people who live close to forest. In fact, community forestry is practiced in south and south-east Asia but it differs from region to region (Wiley, 1977). However, in hot arid regions, there are no closed forests and rural folk practice arable farming in association with multipurpose tree species (MPTs). Thus in true sense, agroforestry is traditional landuse of arid western Rajasthan. Therefore, in hot arid regions, the major issues for extension forestry are as under:

- Forestry in hot arid regions is altogether different from timber forestry.
- Trees on agriculture fields are integrated as part of life support system of rural folk in this part of country.
- Long lived perennial provide both production and protection to fragile ecosystems of the region.

Thus, forestry, commonly known as agroforestry in hot arid regions, in many ways is a private forestry system. Its development approaches are quite different from timber or production forestry. In fact, it is a form of conservation forestry with little benefit of production. However, if such combined productive-protective system is healthy enough, it is capable of providing fuel and fodder supply to farm families the year round (Khan, *et al*. 2006).

The above mentioned concepts target the key areas of forestry extension in hot arid regions. These key issues are very important to design extension strategies especially, in context of arid zone forestry and agroforestry.

Specific Approaches for Forestry Extension in Hot Arid Regions

To respond to the changing needs and objectives of sustainable development in the field of forestry in hot arid regions, it should be backed by: proper education, research and development and extension strategies. For combined protective and productive system like, agroforestry systems of hot arid regions of India, the best assessment of human well beings can be carried out by basic assessment guide (BAG) of Centre for International Forestry Research (CIFOR, 1999a). BAG includes:

- Stakeholder's identification.
- Security of inter-generational access to resources.
- Rights and means to manage equitability.

In addition to provide basic evaluation of stake holders in context of use of tree/shrub products, BAG also provides detailed methodologies for:

- Assessing human well beings in agroforests.
- Ascertaining feelings of closeness that people have to woodlands/agroforests.
- Determining whether or not people have a conservation ethics.
- Assessing production potential of agrofrests/ woodlands.
- Assessing people's voice.
- Analysing people's action affecting forests/ agro-forests/woodlands management.

In fact, BAG is most powerful tool to ascertain stake-holder's identification, security of intergenerational access to resources, rights and means to manage forests/ agroforests cooperatively and equitably, and finally assessing human well beings in totality.

Another important approach for quick forestry extension in hot arid tract is participatory and reflective analytical mapping (PRAM). A society that progress towards sustainability is characterized by a long term trend towards improved environmental conditions by ensuring that the demand for goods and natural resources is as follows:

- It is placed with environmental limits to preclude irreversible environmental degradation
- It actively negotiates among different groups in such a way that their different material and spiritual needs are satisfied.
- From PRAM, the issues vital to forestry extension can be identified in relatively simple way, like
    - What is going to be assessed and why?
    - What the evaluation is trying to achieve?
    - Who is going to gain?
    - How is the evaluation going to benefit?

Thus, PRAM has been found to be useful mode in several ways. It has been recognized as an initial diagnostic tool to design and set up forestry/agroforestry projects for sustainable rural development in hot arid tropics. Moreover, it is an excellent tool in context of periodic evaluation to detect ecological and socio-economic trends. Logical analysis through PRAM is now widely adopted for funding agencies throughout the world (CIFOR, 1999 b). Through PRAM each country or institution or developmental agency or NGO can develop their own forestry related programme according to their regional, social and cultural situations.

Developing Sound Forestry Extension Base for Hot Arid Regions

BAG and PRAM are two most versatile expansion approaches for development of forestry sector. Now the question is, how to develop a sound forestry extension base? Forestry extension system is very weak in the country in general and hot arid regions in particular. The various state forest departments are concentrating on government forest lands. They have no contact with farming communities. For example, in arid western Rajasthan, 80 per cent sand dunes are farmers' owned and they require immediate attention in respect of sand dune fixation but due to lack of relationship between forest department and communities, there is even no road map for this important task. Therefore, the state forest department should develop a separate forestry extension wing. The officers and field level functionaries should be trained through:

- Proper forestry extension education involving educational psychology, principles of teaching and learning, and concepts of social welfare.
- Developing community relation through public relations, community service and community participation.
- Establishing effective communication channels through traditional approach, as well as, through vigorous use of modern electronic media.

Today, the successful forestry extension has only two examples i.e., Joint Forest Management (JFM) and establishment of 'Van Panchayats' by newly created state of Uttaranchal. In these two forest extension systems, forest departments and communities jointly manage natural forests and share responsibility and user rights (Singh, 1996; Anonymous, 2002). However, both the cases are more related to natural forests. The major concern today is the afforestation and reforestation of problematic lands and the forest lands which have been exploited to an alarming extent, respectively. Till date, nothing is said and done for private land forestry, which is of major concern in hot arid tracts. For private land forestry like agroforestry and trees outside forests inter-sectorial approach could be very beneficial. The officers, subject matter specialists and field functionaries of agriculture extension system could be trained for forestry extension, and in turn, in addition to agriculture development, they could develop, design and implement forestry programme as per need of communities or individual farmers. This would also bridge the gap between forestry and agriculture sector. Thus, the communication gap, which today exists throughout the country between agriculture and forestry sector can be overcome by such an inter-sectorial approach of agriculture and forestry extension activities.

Epilogue

Based on our experience in forestry and agroforestry extension in hot arid regions, we found that challenges of assessing human well beings quickly, easily and reliably is not an easy task. This is the first and most important task of forestry extension agencies. To accomplish this important task for hot arid regions, a well accepted problem scoring and analysis guide should be prepared at first instance. The standardization of such scoring guide will make programme implementation much easier. The next step should be the development and implementation of forestry/agroforestry programmes, which are need based according to specific area. A sound forestry extension system based on inter-sectorial approach can do wonders for both forestry and agriculture development in hot arid regions because the farming system in the hot arid region heavily depends on forests/multipurpose trees with crops and animals. A holistic view has to be taken for extension forestry in arid region.

References

Anonymous. 2002. *Van Panchayat Adhiniyam*. State Forest Department of Uttaranchal, Dehra Dun, Uttaranchal.

Anonymous. 2003. *CAZRI Annual Progress Report*. Central Arid Zone Research Institute, Jodhpur, 174p.

CIFOR. 1999 a. *The Grab Bag: Supplementary Methods for Assessing Human Well-Being*. Centre for International Forestry Research, Jakarta, Indonesia, 64 p.

CIFOR. 1999 b. *Scoring and Analysis Guide for Assessing Human Well-Being*. Centre for International Forestry Research, Jakarta, Indonesia, 140 p.

Isager, L.; Theilade, I. and Thomeson, L. 2002. *People's Participation and the Role of Governments in Conservation of Forest Genetic Resources.* Danida Forest Seed Centre, Guidelines and Technical Notes No. 62, Danida Forest Seed Centre, Denmark, 22 p.

Khan, M.A., Tewari, J.C., Singh, Raj and Narain, P. 2006. Structure, production attributes and management strategies in a traditional extensive agroforestry system in an arid region watershed of India. *Forests, Trees and Livelihoods* (U.K.), Vol. 6 (in press).

Narain, P. and Tewari, J.C. 2001. Agroforestry and soil conservation in arid regions of India. *Proc. First Regional Workshop on Thematic Programme Network-2 on Agroforestry and Soil Conservation*, UNCCD; MOEF, Govt. of India; CAZRI, Jodhpur and GM, Govt. of Germany, ICRISAT, Hyderabad, pp. 3-13.

Narain, P. and Tewari J.C. 2005. Trees on agricultural fields: A unique basis of life support in hot arid zone. In: *Multipurpose Trees in Tropics: Management and Improvement Strategies* (eds.) V.P. Tewari and R.L. Srivastava (MOEF, GOI; APAFRI; Goethe Institute, Max Muller Bhawan, New Delhi; DST, GOI; National Medicinal Plant Board, New Delhi), Arid Forest Research Institute, Jodhpur, pp. 115-120.

Pande, S.K. 1995. Sustainable forest management: An Indian perspective. In: *Implementing Sustainable Forest Management in Asia and Pacific* (eds) A. Johnson and P.B. Durst, FAO of UN, Rome, pp. 109-121.

Shekhara, P.C. 2001. *Private Extension: Indian Experience.* National Institute of Agricultural Extension Management, Hyderabad 500 030, 226 p.

Singh, S. 1996. Joint *forest management in India. In:* Decentralization and Bio-diversity Conservation (eds.) E. Lutz and J. Caldecott, The World Bank, Washington DC.

Tewari, J.C.; Tripathi, D.; Narain, P. and Singh, S.P. 2003. A study of the structure, energy fluxes and emerging trends in traditional central Himalayan agroforestry system. *Forests, Trees and Livelihoods* (UK) 13: 17-38.

Wily, L. 1997. *Villagers as Forest Managers and Governments "Learning to Let Go": The Case of Duru-Haitemba & Mgori*

Forests in Tanzania. Forest Participation Series No. 9, International Institute for Environment and Development, forestry and Landuse Programme, Tanzania.

FRAMEWORK FOR THE FUTURE DEVELOPMENT OF EXTENSION ACTIVITIES IN THE FORESTRY SECTOR

V.N. Joshi, S.K. Intodia and R.C. Dadheech

Directorate of Extension Education,
Maharana Pratap University of Agriculture & Technology,
Udaipur (Rajasthan)

ABSTRACT

For maintaining ecological balance, the country need about 33 per cent of the land area under adequate forest cover. But at present forest cover is only 19.5 per cent in India. The rate at which degradation of forestland is taking place is alarming. Rajasthan is one of the forest poor states in the country and forest cover is only about 9 per cent.

The Government of India, State Governments and Union Territories have made sincere efforts for development and protection of existing forest in the country since independence. But unfortunately, the desired success has not been achieved because of lack of people's active participation in various forestry programmes.

Now, a need has been experienced that the people's participation in various afforestation programmes should be considered seriously. People's participation should be ensured in planning, execution, management and protection of forestry programmes.

A strategy should be worked out for future extension activities in the forestry sector. For this purpose, an Integrated Forest Development Programme (IFDP) is to be formulated.

Under this programme, people's participation should be ensured for assessment of their needs through conducting PRA (Participatory Rural Appraisal) in villages for selection of site and type of tree species for any forestry programme. Villagers should be motivated to accept the new responsibility through group discussions, awareness camps, study tours and training programmes. The extension scientists have to evolve suitable strategies for popularization of generated technologies and benefits desired from tree planting materials among rural masses. The Village Forest Protection and Management Committee (VFPMC) should be constituted as registered society and incentive should be provided to the society. The VFPMC after identification of targets of planting of specific forest trees based upon the assessment of agro-ecological situation of the area should be entrusted with the responsibility of planting the forest trees and their subsequent care for increasing the forest area. Further, when the forest plants get developed and they are ready to provide the forest products, then at that time a share in forest products could also be considered for VFPMC for effective management of forest area. In a democratic country like India, we can achieve targets as suggested by National Forest Policy by active participation of rural masses in forestry programmes.

Introduction

For maintaining ecological balance, the country need about 33 per cent of the land area under adequate forest cover (Dwivedi, 1992). But at present forest cover is only 19.5 per cent in India. The rate at which degradation of forestland is taking place is alarming. Rajasthan is one of the forests poor state in the country and in the state, forest cover is only about 9 per cent. This is due to heavy pressure of human population on land, growing demand of timber, fuel wood, fodder and grazing encroachments, shifting cultivation, urbanization, industrialization and improper management of forest and waste land.

About 17% of the worlds' population and 13% of the world's livestock population has concentrated in 2.4% of the earth surface in India. India has the largest wasteland area in the world and 3% of forest is lost annually.

Due to degradation of forest cover over a large area, various adverse effects are evident. It has been continuously reducing the life supporting capacity of the earth and adversely affecting environment. Frequent floods, drought, severe soil erosion, reduced water yield, decreasing productivity and reduced harvest from the forests are some of the consequential adverse effects. Diminishing national resources (forest) is a matter of serious concern.

India is one of the 12 mega-biodiversity areas with 5% of the world's biodiversity having 45,000 species of flora and 65,000 species of fauna. The number of extinct and endangered species is more than 200 species of fauna and 3000 species of flora. The biological impoverishment of the country is a serious threat to bio-diversity of our fragile ecosystem.

Further, the environmental status is being threatened rapidly due to man induced pressure for rapid economic development resulting in drought, famine, flood, siltation, environmental pollution and their impact on natural resources such as land, water, biomass and the future prosperity of the present and succeeding generations. Environmentally hostile externalities such as mining and industries, when improperly managed, tend to endanger and degrade the ecosystem without compensating for it.

Extension Strategies in the Forestry Sector

The Government of India and the State Governments have made sincere efforts in forestry sector since independence. But, the success is not as per expectation. There are several reasons for this. One of the reasons is that rural people considered various forestry programmes as Govt. programme. They did not show keen interest and participation in reforestation of degraded forests. Their participation in forestry programmes was negligible. Once any forestry programme has been over, then people started grazing of animals and cutting of trees from rehabilitated forest area. To overcome this problem, people's active participation should be ensured in forestry programmes. A strategy should be worked out for integrated forest development programme with active participation of villagers. In various programmes of forestry, success has been achieved when people participated in that programme.

1. Integrated Forest Development Programme (IFDP)

(i) Rapport Building

Before starting any forestry programmes in particular area, rapport should be established with villagers by organizing night camps, contacting local leaders/senior citizens/progressive farmers, campaign, film show, etc. Such programmes should be entrusted to SAU's for their effective implementation.

(ii) Training

Both on campus and off campus, trainings should be imparted to rural masses to create awareness and to educate regarding forestry programmes in terms of various technological interventions such as:

(a) Agro-forestry and social forestry to meet the essential needs of local population for fuel wood, fodder, small timber, non-timber forest products etc.

(b) Reclamation of problematic soils (ravines, shifting sand and marshes, swamps and waterlogged soils): In villages barren and problematic soil should be reclaimed. These reclaimed soils could be used for afforestation, agro-forestry, silvi-pastoral purpose as per the classification of land.

(c) Industrial plantation and reclamation of mined area: A sizeable area of forest lands is degraded every year due to mining and industries. These mined areas should be reclaimed with plantation of forest trees.

(d) Integrated watershed develop project: In watershed projects, people could be motivated for forestry in class IV and below lands through trainings.

(e) Construction of water harvesting structure: The success rate of plantation is less because of scarcity of moisture in soil. Rural masses should be motivated for rain water harvesting and construction of water harvesting structure.

(f) Construction of vegetative / stonewall barrier to facilitate natural regeneration.

(g) Reforestation/rehabilitation of degraded forest through supplementary planting of eco-compatible and economically useful species.

(h) Need based site specific treatment of public wastelands for Silvi-pastoral development through intensive planting of socially accepted species.

(i) Various models *viz*., agri-silvi model, agri-horti-silvi model, silvi-pastoral model and homestead planting based on scientific land use and crop production pattern.

(iii) Identification of technological interventions with selection of site and tree species

Need based technological interventions, need based site and socially accepted, and economically important trees species should be identified by using PRA approach in that particular area. If technological intervention and site of forest and tree species are as per their need and demand, then villagers will show their interest in the programme. Sites of technological interventions should be selected in such a way that the regenerated areas ensure benefit and employment to rural youth over a continued time frame.

2. Village Forest Protection and Management Committee (VFPMC)

The Village Forest Protection and Management Committee (VFPMC) can be constituted as registered society. Efforts should be diverted towards providing orientation inputs to the field front line staff and members of VFPMC on participatory appraisal, micro-planning-macro-perspective of integrated sustainable rural resource management, production system analysis, harvesting, processing, value addition and marketing system.

Nursery can be established by VFPMC to provide planting material (grasses, shrubs and tree species) for sowing purpose. Each tehsil should have nursery to meet the requirement of rural people according to choice of species for sowing and ecological compatibility.

Incentive should be provided to rural youth for voluntary work on building group formation at voluntary level for effective protection management and establishing cottage-scale industries at the village level.

It would also be necessary to understand and learn from the existing management practices and their ecological and economic impact in a range of location - specific situation and thereby develop simple technologies for use by grass root village institute, local communities, NGO's and lower level field functionaries of extension/ forest department.

Literature consulted

Dwivedi, A.P. 1992. Agro-forestry- Principles and Practice. Oxford & IBH Publishing Co., New Delhi.

Dwivedi, A.P. and Gupta, G. M. 1992. Afforestation of Arid Lands. Scientific Publishers, Jodhpur.

Rathore, N.S. and Mathur, A.N. Forestry for Arid Regions. Agrotech Publishing Academy, Udaipur.

Shah, S.A. 1988. Forestry for People, ICAR Publication, New Delhi.

Singh, S.P. 1994. Handbook of Agro-forestry. Agro-Tech. Publishing Academy, Udaipur.

Tejwani, K.G. 1994. Agro-forestry in India. Oxford & IBH Publishing Co., New Delhi.

A CONCEPT NOTE ON STRATEGIES AND APPROACHES FOR EXTENSION OF FORESTRY TECHNOLOGY IN ARID ZONE

R.S. Mertia and Rajendra Prasad

Central Arid Zone Research Institute,

Regional Research Station, Post Box-63, Jaisalmer-345 001 (Raj.)

ABSTRACT

Basic elements for planning forestry extension programmes include environmental setting of the area, target groups/ users/beneficiaries and tested technologies. For successful transfer of a selected technology, the socio-economics, needs and willingness of the target group must be analyzed. Most of the target groups look forward for complete package of technology supported with data for its genuineness and benefits before adoption. The future strategies must emphasize on organization and its staffing, linkages with other agency, development of infrastructures and extension material. Extension agencies should have all infrastructure, facilities and well equipped audio-video technical tools. The extension materials in the form of leaflets, pamphlets, posters, charts, small bulletins, films etc help in motivating the people. For making forestry technology acceptable to the people of various target groups in arid zone, water harvesting/ soil moisture conservation techniques, protection of seedling from stray cattle, special incentives in cash or kind, planting of preferred indigenous multipurpose trees, relaxation in legal restrictions put on utilization and felling of tree crop raised by farmers/ tree growers should be given due consideration.

Approaches should include trainer's training to acquaint them with technological packages and communication skills, entry point activities to have first hand assessment of the environment and feasibility, conducting need based training for users, distribution of extension material and on farm trial for demonstration of particular technology. Mass meeting, workshops, symposia and seminar may be organized to address large groups and feedback be accounted for further modification.

INTRODUCTION

Indian arid zone occupies about 32 mha area constituting 12% of the geographical area of the country. It is spread over Rajasthan (61.8%) Gujarat (19.6), Punjab (4.6%), Haryana (4.0%), Andhra Pradesh (6.8%), Karnataka (2.8%) and Maharashtra (0.4%). The region is almost devoid of close canopy forest and area under forest is very low (2.45%). The area under forest in Rajasthan, Gujarat and Punjab & Haryana is only 2.61, 2.31and 1.35 per cent, respectively. The per capita forest in the region is 0.04 ha against national average of 0.11 ha and world average of 0.8 ha. The existing forests are highly degraded and productivity is very low (0.05 cu m /ha/year) as compared to national average of 0.7 and world average of 3.10 cu m/ ha/year. Hegde, *et al.* (1989) reported that India has lost three quarter of its forest in three quarter of a century and acute shortage of fuel, fodder, timber, bamboo and other forestry produce is witnessed in the country. In Rajasthan, the gap between demand and supply of fuel and fodder is estimated to the tune of 1.65 and 16.0 million tones per annum, respectively. The increasing gap between demand and supply of the forest products and services invite excessive biotic pressure on existing forestland, which accelerate the process of degradation. In the given circumstances, dissemination and extension of forestry technologies to the end users is the only option to save ever declining forest resources and generate new avenues for meeting basic needs of the people. The present paper deals with the concept, strategies and approaches of forestry extension education with special reference to arid zone. The basic objective is to suggest orientation of forestry extension programmes for better acceptability and adoptability.

BASIC ELEMENTS FOR PLANNING FORESTRY EXTENSION PROGRAMMES

Environmental setting

The environmental settings of the area is the launching pad for forestry technology and hence, it is essential to have basic information such as land physiography, climate (rain fall pattern), land use pattern about the area and availability of surface, as well as, ground water resource. This would help in defining and focusing objectivity and relevance of forestry technology aimed for extension. While planning forestry extension programmes for arid region of Rajasthan, the climatic constraints need to be the guiding spirit for selecting particular technology.

Target groups / users / beneficiaries

The principal users of forestry technology remain the state forest departments, however, the issues like desertification, land degradation, low productivity, pastoral economic base of the habitants and occurrence of frequent droughts in arid region attract all sects of people forming different target groups with varying interests. A farmer would plant trees with the objective of fuel, fodder and small timber while an environmentalist may admire it for its role in balancing the ecological system. A teacher may be interested in providing shade to its school campus while forester would look toward its timber value. It is really challenging for the extension agency to motivate all these target groups for advancement of forest / tree cover area. Further advances in forestry research may attract few more user groups in future, which would increase horizons of forestry extension programmes. Long-term strategies of extension must be conceived in such a way that they would keep pace with changing scenario. To increase the acceptability of particular technology, it would be proper to identify right group of people in given set of agro climatic condition.

For successful transfer of a selected technology, three basic elements *viz.* socio-economics, needs and willingness of the target group must be analyzed before taking voyage of specific technology to them. This would give a fair idea about the practical feasibility and expected adoption by the users.

Tested technologies

This is needless to emphasize that only tested technologies should be taken to the users for their adoption. Most of the target groups look following elements before adopting any new innovation as claimed by the extension agency.

- Complete package of technology supported with data for its genuineness and benefits
- Practical feasibility
- User friendly
- Attractive innovation

While preparing technological package, following question must be answered by the technologists:

- Whether the technology fulfills the need or technological gaps?
- What is optimum level of inputs required?
- What will be yield in physical and monetary term?
- Is processing, handling and marketing of the produce easy?
- If above questions remain unanswered, the successive adoption would embrace an element of doubt.

STRATEGIES

Organization and its staffing

The staff required for implementing of forestry extension programmes must be different than those engaged in general or production forestry. The extension staff should be technically sound and have correct attitude toward the programme. They should have the social skill and ability to communicate with local people in concept, words and idioms they understand. In addition to structural improvement, the functioning of the extension organization must be modified to keep pace with changing social environment.

Linkages

Good linkages between all government organizations, NGOs, self help groups are essentially required. It will be ideal if forestry extension programmes are implemented in close coordination of other developmental programmes for rural communities.

Infrastructures and extension material

Extension agencies should have all infrastructure, facilities and well equipped audio-video technical tools. The extension materials in the forms of leaflets, pamphlets, posters, charts, small bulletins, films etc. help in motivating the people.

Basic consideration in arid zone

Scarcity of water, uncontrolled grazing, pastoral economy and fragile ecosystem make the success of tree plantation difficult in arid area. For making forestry technology acceptable to the people of various target groups in arid zone, following points are to be emphasized (Prasad, 1998).

- Water harvesting / soil moisture conservation techniques should be focused, as the main causal factor for mortality of seedlings is scarcity of water.

- People should be motivated and taught to protect seedlings from stray cattle.
- Special incentives in cash or kind should be given to farmers / tree growers.
- For small farmers and landless people, home yard planting should be encouraged.
- Planting of preferred indigenous multipurpose trees, which are well adapted to the climatic condition, needs to be encouraged.
- Legal restrictions put on utilization and felling of tree crop raised by farmers / tree growers should be removed.

APPROACHES

1. Trainers training

The extension staff must have sound knowledge of the scientific package that they are taking to the users. Besides, they should be able to communicate effectively with the users while advising users to adopt certain package of practices or technology. To develop a team of extension staff with positive attitude, the illustrative foundation training should invariably follow capacity building trainings. The periodic training be conducted for updating knowledge and parallel development. To enrich the scientific knowledge base and communication skills of the extension staff, they should be imparted in-depth training on objective oriented specific subject areas. For this purpose the researchers / scientists who have generated or developed particular technology be honored as resource person for expert opinion.

2. Entry Point Activity (ETP)

This ice breaking activity between the extension agency and target groups facilitates in building rapport and assessing environmental feasibility for any programme. At this stage programme must be made transparent and explained to the beneficiaries. Extension team must be cautious at this stage on the provision of the subsidies and facilities to be provided of OFTs or demonstrations. Nothing should be promised beyond schedule of the programme. Many times false promises in beginning prove to be fatal for success of the programmes. Diction of the people reflecting ground level realities to be honored and taken as guiding for the programme.

3. Users training

Once having decided an area for grounding of transfer of technology programme, first task of extension team is to impart training to identify elite innovators / respondents/ beneficiaries to inject an idea for reverviration in the society. Both, on campus and off campus trainings should be conceived keeping in mind the target groups and technology to be transferred.

4. Extension Material

Appropriate extension material in the form of published literature (brochure, bulletin, leaflets, pamphlets posters, charts etc. and audio and video CDs) on transferable technologies is the pre-requisite for motivating respondent for adoption. This material should be shown and distributed to potential users. Use of extension material judiciously and timely in proper place is as important as its development. As far as possible, published extension material should be made in easily understandable local language with sufficient illustrations.

5. On Farm Trial (OFT)

Demonstration of any technology by conducting OFTs is one of the most successful methods of extension. Sites of OFTs should be selected after having analyzed all the potential and constraints of the area and target groups. Wrong site selection may cause failure of any proven technology and once it has happened the confidence of user is lost and whole efforts of the extension agency goes waste.

6. Mass Addressal

Mass meetings, workshops, seminars, symposia, farmer's fairs may be organized to make a point before a large group of people. The feed back from discussion be accounted for further modification and refinement. Use of press, TV and AIR for creating general awareness about forestry plantations and their role in protecting environment widen horizon of audience. Hence these media must be explored for popularizing certain technologies.

References

Hegde, N.G., Relwani L.L and Kelkar, V.D. 1989. Promotion of fodder and fuelwood trees. BAIF Development Research Foundation Pune.

Prasad, R. 1998. Forestry Extension: Strategy to make forestry technology acceptable to the people. *Indian Forester*, 124 (11): 939-944.

STRATEGIES FOR EXTENSION OF IMPORTANT NON-WOOD PRODUCTS TECHNOLOGIES

Ranjana Arya, Mala Rathore, S. H. Jain, Hemant Kumar, R.R. Lohara and R.K. Meena

Non Wood Forest Products Division,
Arid Forest Research Institute, Jodhpur 342 001 (Raj.)

ABSTRACT

Increased concern over the degradation and loss of forests has brought attention to NWFPs. They are increasingly being seen as offering new alternatives to timber extraction, forest conversion and destructive forms of forest exploitation. Improved govt extension services for non-wood forest products through such measures as technical assistance, credit and market opportunities will directly benefit people.

Technology for rehabilitation of barren degraded salt affected areas of arid Rajasthan is developed for fodder production. Mounding techniques like double ridge mound and circular dished mounds enhances the survival of any plant species under waterlogged condition. This work is taken up for the first time in India. Growth and biomass production of indigenous, as well, as exotic plant species *viz Salvadora persica* (khara jaal), *Azadirachta indica* (neem), *Acacia ampliceps*, *Acacia colei*, shrubs of genus *Atriplex*, *Sueda nudiflora* (unt morad) is enhanced multifold by the use of gypsum, farm yard manure and nitrogen.

Crude protein contents of various famine food (normally not utilized as food) when determined, khejari bark, bui leaves and kumat seed were found to contain 23, 32 and 36% crude protein content respectively indicating their utilization in famine times.

Methodology for preparation of Leaf protein concentrates of eleven important forestry tree species (especially neem and moringa) was developed.

Consumption of fats and oils derivatives in non-food use has increased as consumers incline towards more natural and environment friendly products. Non-edible oil of tree origin is finding increasing applications in cosmetics, medicine, soap manufacturing, flavoring and perfume industries. Seed oil content of important plant species *viz. Salvadora oleoides, S. persica, Balanites aegyptiaca, A. indica, Pongamia pinnata, Cucumis callous, Citrullus colocynthesis* of arid region have been determined.

Survey of Sandal population in south (Udaipur, Banswara, Jhalawar, Sirohi and Pali districts) and eastern part (Jaipur, Karoli and Ajmer) of Rajasthan indicated presence of natural stands of sandal with good potential for its development in future. Heartwood and oil content varies with place and age and the oil content ranges from 0.9 to 3%.

There is a need to extend these technologies to end-users. As a first step, appropriate end users need to be identified *viz* SFD's, NGOs, Govt Departments, forest product based industries, farmers etc. Local and dedicated extension worker will be the most important link between technology developed and its reach to the end user. Proper incentive should be given to achieve excellence. Proper co-ordination between end user, extension worker and the parent department is essential. Demonstration of technology through field & laboratory visits, trainings, modern audio-visual techniques (films, radio & TV talks, books, journals, bulletins, newsletters, handouts, posters, hand bills etc), intensive training to field staff, lecture, seminars, symposium and farmer fairs are the most important tools for proper extension.

Introduction

In recent years, non-wood forest products (NWFPs) have attaracted considerable global interest. This is due to increasing recognition that NWFPs can provide important community needs for improved rural livelihood; contribute to household food security and nutrition; help to generate additional employment and income; offer opportunities for processing enterprises; contribute to foreign exchange earnings and support biodiversity conservation and other environmental objectives. NWFPs are also often called Non-Timber Forest Products (NTFPs). They are also called "extractive", "secondary" and "minor" forest products. The term 'non-timber forest product' encompasses all biological materials other than timber which are extracted from forests for human use. Among the most important NWFP regarding their value in international trade are medicinal plants (US$ 689.9 million), nuts (593.1), ginseng roots (389.3), cork and cork products (328.8), and essential oils (312.5) (Walters, 2002; Iqbal, 1995)).

Non-Timber Forest Products include

Edible plant products: Food, edible oil, spices, fodder, other edible plants.

Non edible plant products: Rattan, bamboo, sustainable produced wood, ornamental plants, chemical components, other non edible plant products.

Medicinal Products

Work done at Arid Forest Research Institute

Non wood Forest Product division, AFRI was established in 1992 with the main aim of doing phyto-chemical investigation on arid zone flora for exploring their utilization. Research findings on some aspects of non-wood forest products carried out at Arid Forest Research Institute are given below:

Fodder production from salt affected areas

Wastelands statistics indicate that about 63.85 million hectares, which account for 20.17% of the total geographical area (328.72 million hectares), exist as wastelands in India. Very high percentage of area under wastelands, about 31% mostly sandy and about 22% in mostly scrub respectively occurs in the states of Rajasthan and Gujarat. Approximately 1.04 m ha salt affected wastelands are reported in states of Rajasthan and Gujarat. There are two major tracts in arid western Rajasthan having exceptionally large and wide spread aerial occurrence of soil salinity. One is the area of far flood plain deposits of Ghaggar system in the north-west, adjoining Haryana State and the other is in the south-east identified broadly as Bilara-Sumerpur-Jalor triangle. Experimental trials were laid at Gangani in Jodhpur, Rajasthan to find out species, which can produce fodder; fuel and other minor produce and suitable planting practice packages to make these lands productive.

Salient findings are as follows

Exotic shrubs of genus *Atriplex* are known to produce nitrogen rich fodder used for sheep and goat. Fodder has high salt content and hence is mixed with cereal residue (Runciaman and Malcom, 1989; Abou El Nasr *et al.*, 1996). They perform well on arid saline alkali lands with FYM and nitrogen. Application of nitrogen increased the growth of *Atriplex lentiformis*. CAN (Calcium ammonium nitrate) was found to be better source of nitrogen than Urea. The green biomass yield was ranging from 1 t ha^{-1} to 3.2 t ha^{-1} in different treatments (Anon, 2001).

Salvadora persica (L.) is a preferential halophyte, evergreen multipurpose tree. However, its slow growing nature resists its cultivation. Application of nitrogen in combination with gypsum gave better results for growth and biomass as compared to application of nitrogen only. T_6 (gypsum + 9g N) was the best treatment attaining 207 cm of height and 212 cm of crown diameter and 12.0 kg/tree of above ground fresh biomass, which was 38%, 24% and 71% more than the untreated plants (Arya and Lohara, 2004). However, maximum leaf fodder, 2.9 t/ha was recorded in T_8

treatment (Gypsum + 27 g N) and was 41.5 % more than the control (2.0 t/ha, Table 1).

Table 1. Total biomass and leaf fodder production from *Salvadora persica* (t ha^{-1}) with different treatments at six years of age

	Total Green Mass	Green Leaf Mass	Total Dry Mass	Dry Leaf Mass
T_1	5.76	2.05	2.57	0.68
T_2	7.32	2.20	3.54	0.72
T_3	5.79	1.77	2.69	0.57
T_4	6.09	2.15	2.79	0.65
T_5	6.57	2.27	3.05	0.78
T_6	9.83	2.78	4.72	0.94
T_7	6.27	2.38	2.79	0.72
T_8	9.02	2.90	4.22	1.00

$T_1 = G_0 N_0$, $T_2 = G_0 N_1$; $T_3 = G_0 N_2$; $T_4 = G_0N_3$; $T_5 = G_1 N_0$, $T_6 = G_1 N_1$, $T_7 = G_1 N_2$; $T_8 = G_1N_3$

Double ridge mound (DRM, adapted from Australia) and circular dished mound (CDM) technique appreciably increased the survival by providing protection from waterlogging and less salty environment. Less salt tolerant species *viz.* *Acacia colie* and *Colophospermum mopane* (exotic), *Azadirachta indica* and *Zizyphus mauritiana* (indigenous) maintained very high initial survival.

Sueda nudiflora and *Atriplex stocksii,* are perennial shrub species found on mud flats or in saline soils in Gujarat and used as fodder. *S. nudiflora* recorded nearly 100% survival, attaining max. growth Overall DRM was the best planting practice producing maximum 1.49 t ha^{-1} mean dry biomass, followed by CDM (1.42 t ha^{-1}) and control (0.53 tha^{-1}) for shrub species, *Sueda nudiflora, Atriplex stocksii* and *Atriplex lentiformis* (Arya *et al.*, 2006). The mean yield of dry leaf fodder was also highest on DRM (0.70 t ha^{-1}) followed by CDM (0.57 t ha^{-1}) and control (0.22 t ha^{-1}).

Famine Food and Leaf protein concentrates as food supplement for utilization during drought

In arid and semi-arid areas where forest cover is very less, trees outside the forest can in general make valuable contributions to food security and provide income to the rural poor. Protein and calorie malnutrition is widespread in these areas leading to high child mortality rate. Crude protein (CP) contents of seeds, leaves and branches of locally used famine foods of arid region have been determined. CP of bark of *Acacia leucophloea, Acacia nilotica* and *Prosopis cineraria* has been found to be 18.9%, 13.7% and 23.1% respectively. Leaves of *Cleome viscosa, Aerva javanica, Cassia tora* and *Balanites aegyptiaca* contain 29.9%, 31.9%, 29.3% and 26.3% protein respectively. *Acacia* seeds contain maximum amount of CP *viz. Acacia senegal* has 35.8% protein. *Capparis* and *Prosopis* form a good nutritious source of food. They can be called famine foods attracting additional consumer category.

Trees Outside Forest can be used as a source of leaf protein concentrate (LPC), a highly nutritious food. LPC should be seriously considered as additional protein source in the case of non-ruminants and man, especially, in drought prone area. Methodology was standardized to determine the potential of arid zone plants for preparation of LPC. Extraction characteristics of eleven tree species have been studied and the quality of LPC prepared from them was investigated. Maximum plants were found to belong to type II category. Different fractions, chloroplastic and cytoplasmic proteins, were analyzed for their crude protein contents. Analysis of LPC showed considerable difference in their protein contents, ranging from 10.9% to 76.4%. Based on these results the trees *Moringa oleifera* and *Azadirachta indica* were found best suited for LPC preparation. Incorporating LPC, as supplement in the mid-day mea scheme run by the Govt, may be useful in improving nutritional status of children in drought hit Rajasthan (Rathore and Meena, 2004).

Table 2. Crude protein content of some arid zone species from Rajasthan used in famine

Species	Part used	Crude protein (%)	Species	Part used	Crude protein (%)
Acacia nilotica	Pods	13.1	*Citrullus colocynthis*	Seed	28.3
Acacia nilotica	Seed	32.8	*Cleome viscosa*	Leaves	29.9
Acacia nilotica	Bark	13.7	*Euphorbia caducifolia*	Leaves	26.3
Acacia senegal	Pods	22.0	*Indigofera linifolia*	Seed	32.3
Acacia senegal	Seeds	35.8	*Prosopis cineraria*	Seed	35.8
Aerva javanica	Leaf	31.9	*Prosopis cineraria*	Pods	23.2
Achyranthus aspera	Leaves	19.2	*Prosopis cineraria*	Bark	23.1
Balanites aegyptiaca	Fruit pulp	19.2	*Prosopis cineraria*	Leaves	23.2
Balanites aegyptiaca	Seed	27.5	*Tribulus terrestris*	Leaves	30.0
Calligonum polygonoides	Stem	23.5	*Zizyphus mauritiana*	Seed	11.5
Calligonum polygonoides	Flower	13.3	*Capparis decidua*	Fruit	24.2
Cassia tora	Leaves	29.3	*Capparis decidua*	Seed	30.5

Source: Rathore and Meena, 2004

Potential oil yield from non-edible oilseed species

Tree borne oilseeds are the potential source of fatty oil, which constitutes the major industrial raw material for the manufacture of many value added products *viz.* soap, cosmetics, medicines, lubricants, oleo-chemical etc. Seeds of *Salvadora oleoides* and *S. persica* collected from different areas of Rajasthan and Gujarat were investigated for their oil content and these were found to have maximum oilcontent from Jaipur (44.2%) and Udaipur (45.3%) respectively (Tripathi, *et al.*, 1996) in Rajasthan and from Bhavnagar and Mehsana in Gujarat. *Azadirachta indica* seeds analysed from Banswara region have the highest oil content 43.2% whereas seeds from Jaisalmer have the lowest 32.4% oil content. Maximum and minimum oil content, 41.85% and 30.7% was found in Seeds of *Pongamia pinnata* collected from Dungarpur and Alwar, respectively. General screening of seeds of arid zone species revealed that *Balanites aegyptiaca, Cucumis callosus Citrullus colocynthis, Tephrosia villosa* & *Tecomella undulata* seeds have 40.7%, 20.3%, 20.9%, 15.7% & 18.0% oil content.

Potential of Sandal in Rajasthan

Recent survey of Sandal population in Udaipur, Rajsamand, Chittorgarh, Pratapgarh, Pali, Sirohi, Banswara, Dungarpur, Jhalawar, Ajmer and Karoli forest divisions of Rajasthan indicated presence of natural stands of sandal with good potential for its development in future. In India, yields of about 0.9 per cent oil have been reported from the heartwood of 10-year old trees, while mature trees of 30-50 years age have yielded 4 percent oil. The oil content in trees of Rajasthan varies between 0.9 to 3% (Table 3). Trees grown on farmland with irrigation posses less heartwood and oil, whereas naturally well grown trees contain more heartwood and oil. Heartwood and oil content found increases with age and girth in the arid region (CAZRI campus Jodhpur) showing that this high value tree species can be grown as an agroforestry crop in arid and semiarid areas of Rajasthan.

Table 3. The oil content in trees of Rajasthan

Location	GBH (cm)		Expected Heart-wood yield (Kg)		Oil yield (%)	
	Min	Max	Min	Max	Min	Max
Jodhpur (CAZRI, Campus)	45	55	12.66	24.83	0.95	3.08
Pali	43	67	10.87	48.16	0.94	2.29
Udaipur	57	92	27.99	139.60	0.14	3.07
Chittorgarh	39	60	7.83	33.25	0.98	2.77
Pratapgarh	28	36	All sap	5.99	nil	1.18
Dungarpur	37	62	All sap	37.12	nil	2.44
Jaipur	50	80	18.03	87.33	1.10	2.09
Karoli	59	95	31.43	155.47	1.74	3.15
Ajmer	57	67	27.99	48.16	2.26	1.61
Sirohi	55	70	24.83	55.78	2.59	0.88
Banswara	90	92	129.67	139.60	3.29	0.99
Jhalawar	51	57	19.27	27.99	1.97	2.37

Source: Jain, *et al.*, 2004

General constrains and recommendations for sustainable NWFP Production

Constraint 1: There is a profound lack of information necessary to realize the full benefits of NWFPs for individual, community and national well-being; decision-makers, forest managers and resource users alike lack information about economic, ecological and social characteristics of NWFPs and their uses.**Recommendations**

- Govt efforts be strengthened to conduct research and generate, compile and disseminate information and statistics to key stakeholders on NWFP resources and their socioeconomic and ecological values.

- Governments and development agencies support education and public awareness programs for NWFP conservation and sustainable use.

Constrain 2: Lack of protected rights to access and benefit from NWFP resources can adversely affect their conservation and sustainable use and discourage investment in the resource

Recommendation

- Governments, with assistance from concerned agencies and organizations, develop and implement policies and legislation to provide secure access and benefits to the people whose livelihoods are dependent on or supplemented by NWFP.
- Governments, with assistance from concerned agencies and organizations, ensure that stake-holders, particularly collectors, growers and traders are provided incentives to sustainably manage NWFP resources. **Constrain 3:** Individuals, communities

and institutions generally lack the technical, financial, political and social capacity to influence policies and generate information nece-ssary to manage and monitor NWFP resources effectively.

Recommendation

- Governments, with assistance from concerned agencies and organizations, support programs and projects to build individual, institutional, and community-based capacity to manage NWFPs through multi-stakeholder participation.
- Governments, and research agencies give priority to research and the development and dissemination of management practices to be integrated into multi-purpose forest and agro forest resource management. (FAO, 1995 a &b).

Strategies for extension

Transfer of technology approach: We will require the development of new technology if we are to shift to more biologically and economically sustainable systems. We will also need the willingness of the rural community to accept new technology. One approach to promoting technology is by presenting lectures, seminars, farm walk, demonstrations and research station field day talk to groups. This is transfer of technology approach where we extend our ideas and packages of practice to clients.

We have traditionally applied the linear extension model

Research → Knowledge → Transfer → Adoption → Diffusion

Summarized by Russell, *et al.,* 1989 as generally the best suited for extension.

JFM as an approach for extension

NWFPs have been a great source of sustenance to the people living in and around forests during pinch season and help them to tide over the crisis of survival. Tribal populations depend heavily on NWFPs for income and subsistence. Therefore, the Non-wood forest products plays important role in JFM efforts for a number of reasons. First, NWFPs are integral to the lifestyle of forest-dependent communities. They fulfill basic requirements, provide gainful employment during lean periods and supplement incomes from agriculture and wage labour. Medicinal plants have an important role in rural health (Prasad and Bhatnagar, 1991). In parts of West Bengal, communities derive as much as 17 per cent of their annual household income from NWFP collection and sale (Malhotra *et al.*, 1991). According to J.Y. Campbell (1988, cited in Tewari and Campbell, 1995), small-scale forest-based enterprises, many of which rely on NWFPs, provide up to 50 percent of the income for about 25 percent of India's rural labour force.

Second, NWFPs have a decided advantage over timber in terms of the time needed to achieve significant volumes of commercially valuable production. Timber production is a long-term endeavour, and in many areas timber harvesting may not be ecologically desirable. Moreover, many NWFPs become available even in the earliest stages of rehabilitation of degraded forest areas.

Third, at the national level over 50 per cent of forest revenue and about 70 per cent of forest export revenue comes from NWFPs, mostly from unprocessed and raw forms (Tewari and Campbell 1997; Prasad, Shukla and Bhatnagar, 1996).

NGOs are increasingly getting involved with JFM. They have an important role to play in capacity building, information dissemination, monitoring and evaluation, policy making and as facilitators in the programme implementation. However, NGO involvement needs to be on a more continuous and sustained basis, which is however constrained by lack of implementation freedom owing to dependence on donors. NGOs need to work more closely for the implementation of projects that will help to alleviate the pressure on forests, as well as, increase their reach in motivation and community mobilization exercises.

Integrating local knowledge

Since local people and groups often know more about specific NWFPs than many formal scientific institutions, integration of local knowledge with institutional research results improves the effectiveness of extension. Participatory research and extension help to ensure that the empirical local knowledge built over many years of living with the forest is captured and used. In addition, scientific methods can be applied to systematize and expand local knowledge. Participatory extension can help build methodological capacity at the local level for critical analysis of trends and action. The combination of the knowledge resources of the various parties facilitates innovation.

A participatory study on the role of NWFPs in a conservation and development strategy in Pará, the eastern-most state of the Brazilian Amazon, provides an example (Shanley, 1999). Attempts were made to make the data collected as accessible as possible to local people, for instance through the use of or conversion to local units of measurement, through workshops with local people and through song and theatre, in order to facilitate a dialogue with local forest users. This effort also served to break down the barriers between research and users (including extensionists). The extension team developed illustrated booklets presenting the ecological and market data, posters, songs and lore used in workshops, which could be used to reinforce outreach efforts, could serve as a training tool for extensionists and could reach distant communities beyond the team's travel.

Networking

Another approach has been to emphasize networking, especially between technicians working on NWFPs and potential producers and markets. Since technical knowledge about an individual NWFP may be held by disparate partners, technical networking may be an effective way of combining and systematizing technical knowledge and making it more widely available. An example is the work carried out by FAO's NWFP programme.

Networks are usually viewed as a flexible and often informal means of improving information exchange. Although they may cover wide areas of expertise. NWFP networks often focus on rather specific topics.

With the advent of Internet facilities and e-mail, which allow simultaneous contact with a large number of people at low cost, the capacity and efficiency of some networks have been greatly enhanced. Electronic newsletters (e.g. ntfp-biocultural digest, an e-mail list for announcements and facilitated discussions on topics related to non-timber forest products [NTFPs]; Tropenbos NTFP mailing list) and e-mail conferences are effective and cheap tools for exchange of views and animated discussions on specific topics (Anderson, *et al.*, 1999).

Conclusion

Extension has a key role to play in terms of disseminating ideas and technologies developed. There are many strategies available as discussed above. Local dedicated and progressive extension worker will be the most important link between technology developed and its reach to the end user. Proper incentive should be given to achieve excellence. Proper co ordination between end user, extension worker and the parent department is essential. Demonstration of technology through field and laboratory visits, trainings, modern audio-visual techniques (films, radio and TV talks, book, journals, bulletins, newsletter, handouts, posters, hand bills etc.) in vernacular language will greatly improve extension.

References

Abou El Nasr, H.M., H.M. Kandil, A. El Kerdawy, Dawlat, H.S. Khamis and H.M. El-Shaer 1996. Value of processed saltbush and *Acacia* shrubs as sheep fodders under the arid conditions of Egypt. *Small Ruminant Res.*, 24: 15-20.

Anderson, J., K. Warner, L. Russo and H. Qwist-Hoffmann 1999. The challenges of extension for non-wood forest products. FAO, Unasylva 198(3) pg 71.

Anon, 2001. Screening of exotic and indigenous plant species for their performance on salt affected soil with various management practices. Concluding Report, Arid Forest Research Institute, Jodhpur.

Arya, Ranjana, Chaudhary, K. R. and Lohara, R. R. 2006. Studies on mound practices for establishment and growth of various plant species on saline and waterlogged soil in hot arid zone of India. *Indian Forester* (in press).

Arya, Ranjana and Lohara, R. R. 2004. Growth And Biomass Yield From *Salvadora Persica* with Management Practices on Salt

Affected Soil Under Hot Arid Conditions in India. IUFRO International conference on Multipurpose trees growth, assessment and management. 22nd-24th Nov., 2004, AFRI, Jodhpur in Proceedings 2005: pp 102-108.

FAO 1995a. Report of the International Expert Consultation on Non-Wood Forest Products. Non-Wood Forest Products No. 3. Rome.

FAO 1995b. Non-wood forest products for rural income and sustainable forestry. Non-Wood Forest Products No. 7. Rome.

Iqbal, M. 1995. Trade restrictions affecting international trade in non-wood forest products. Non-Wood Forest Products Series No. 8; FAO, Rome

Jain, S.H.; Arya, Ranjana; Kumar, Hemant and Chaudhuri, K. K. 2004. Sandal (*Santalum album* L.) a high value tree (HVT) species - potential multi purpose tree of arid/semi arid region Multipurpose trees in the tropics: Management and improvement strategies. 22nd-24th Nov., 2004, AFRI, Jodhpur in Proceedings 2005: pp 387-392.

Malhotra, K.C., Deb, D., Dutta, M., Vasula, T.S., Yadav, G. and Adhikari, M. 1991. *Role of non-timber forest produce in village economy. A household survey in Jamboni Range, Midnapore, West Bengal.* Calcutta, India, Indian Institute of Biosocial Research and Development (IBRAD).

Prasad, R. and Bhatnagar, P. 1991. Wild edible products in the forests of Madhya Pradesh. *Journal of Tropical Forestry,* 7(3): 210-218.

Prasad, R., Shukla, P.K. and Bhatnagar, P. 1996. *Leaves from the forest: a case study of tendu leaves in Madhya Pradesh.* Jabalpur, Madhya Pradesh, India, Centre for Environment and Sustainable Development.

Tewari, D.D. & Campbell, J.Y. 1995. Developing and sustaining non-timber forest products: some policy issues and concerns with special reference to India. *Journal of Sustainable Forestry,* 3(1): 53-77.

Tewari, D.D. and Campbell, J.Y. 1997. Economics of non-timber forest products. *In* J.M. Kerr, D.K. Marothia, K. Singh, C. Ramaswamy & W.B. Bentley, eds. *Natural resource economics - theory and application.* New Delhi, India, Oxford and IBH.

Tripathi, Y.C., Prabhu, V.V. and Kumar, Hemant. 1996. Oil content of seeds of *Salvadora persica* and *S. oleoides* from different agroclimatic zones of Rajasthan. *Bulletin of Pure and Applied Sciences*, 15B (2): 125-127.

Rathore, M and Meena, R.K. 2004. Nutritional evaluation of some famine foods of Rajasthan desert. *Indian Forester*, 130(3): 304-312.

Runciman, H.V. and Malcolm, C.V. 1989. *Forage shrubs and grasses for revegetating saltland.* Bulletin No. 4153. Department of Agriculture Western Australia.

Russell, D.B., Ison, R.L., Gamble, D.R and William, R.K. 1989. A Critical Review of Rural Extension and Preactice. Faculty of Agriculture and Rural Development, University of Western Sydney, Hawkesbury

Shanley, P. 1999. Extending ecological research to meet local needs: A case from Brazil. *In* T. Sunderland, L. dark & P. Vantomme, eds. *The non-wood forest products of Central Africa - current research issues and prospects for development and conservation.* Rome, FAO (in press).

Walter, Sven. 2002. Certification and benefit-sharing mechanisms in the field of non-wood forest products – an overview Medicinal Plant Conservation, Volume 8, Newsletter of the IUCN Species Survival Commission, Medicinal Plant Specialist Group. Bonn.

STATUS VERSUS FUTURE OF FORESTRY RESEARCH EXTENSION

C.S. Dange, Sunil Kumar, B.M. Dimri and A.K. Chattopadhyay

Agroforestry & Extension Division,
Arid Forest Research Institute, Jodhpur-342005

ABSTRACT

The forestry research threshold technologies dissemination equip and enhance stakeholders knowledge, skill and access to forestry and allied subjects for promotion of their agro-ecological, economic, social and cultural well-being. Despite systematic concerted effort, the productive forestry has not been able to witness much progress vis-à-vis agriculture or in tandem with agriculture to the required extent. In usual practice, short-term research experiments/ trials/ projects remain in offing for long rotation silviculture crops, whereas, the forestry research extension has a long gestation period. The other hassle has been lack of continuity in effective interaction amongst researchers on the one hand and meaningful linkages between researchers and stakeholders on the other that need to come in synergist mode with public sector, private sector and masses in tandem for future extension plans and it's in principle implementation.

Future forestry research technology strategy necessitates applying institutional integrated approach in a coordinated manner through nodal extension wing by establishing and developing prototype of suitable site-specific collaborative models in the demonstrative plots of stakeholders via long-term projects and diversification schemes delivering perpetual remunerative benefits. Also an up-to-date sound resource base of Forestry Research and Development through ultra-modern Library Information System Extension Centre will be essential to disseminate available technical know-how with inbuilt institutional, national and global technological inputs to this common platform. Dissemination of standardized forestry research pursuits for development of forest based agro-ecological resources shall be able to lead to optimum land use productivity enhancement harnessing for sustainable profitability and increase in purchasing power parity.

Introduction

Primarily, subsistence agriculture and animal husbandry are the major occupations of most of the people of our mandated areas of Rajasthan, Gujarat and Dadra & Nagar Haveli. Apart from compensating farmers for poor fertility and pecuniary benefits, and hence low carrying capacity and agricultural productivity of the land in this region, they act as an effective tool to fight severe and frequent droughts. The soil has poor nutrient status. Inadequate nutrient cycling has heavily degraded the landscape bereft of vegetation to further deteriorate the natural regeneration. Hence, vegetation is sparsely distributed. Stray goat, sheep, cattle and human being exert tremendous biotic pressure negating social fencing. Removal of vegetation for fuel wood, fodder and timber for own use and sale is quite common to sustain livelihood. Planning and development of pasture lands to meet the forage and fodder requirement for huge livestock is lacking. Community forestry lands are also under biotic pressure and beset with encroachments. Depletion of ground water and poor recharge mainly dependent on erratic rainfall has caused widespread hardship to the rural population. This also adds to inadequate irrigation facility coupled with conspicuous absence of perennial rivers. Scanty rainfalls result into dry catchments dotted with sparse vegetation. The Sun shines very hot during the day raising the temperature to 52°C in summer season making desert soil and Aravali Ranges' sandstone unbearable for life processes. The nights are comparatively cooler bringing down the temperature to minus 3°C in winters. The day evapo-transpiration rate is also comparatively very high. As a result, the oxygen level and relative humidity is comparatively low. All this keeps people restless and wanting to give extraordinary end results in medium soils. Scarce resources warrant us to maximize output with optimum input. Therefore, herculean efforts are required to give extra-ordinary results in arid and semi-arid areas of Northwest India.

To meet this requirement, much work needs to be done on forestry research extension information dissemination on regular basis and development of prototype of suitable multi-tier site-specific economically viable remunerative research extension demonstrative role models for diversification towards rehabilitation of degraded semi-arid and arid areas through which people's livelihood concerns shall be addressed with suitable land use sustainable development and productivity enhancement with profitability across the globe in order to increase purchasing power parity of the general populace.

Target Groups

The Farmers, Prospective tree growers, Medi-culturists, Local MFP Collectors, Rural Women, Unemployed youth, Traders, Pansaris/Shopkeepers, Testing Laboratories, Herbal-Pharmaceutical-Nutraceutical Manufacturing Units, Practitioners, Non-Governmental Organisations, Environmentalists, N.T.F.P. based Industries, Ayurveda Department, Forest & Wildlife Department, other concerned Departments, other government and non-government Institutions/Organisations, School/College/University Students, Academicians, Researchers, Scientists, Local Government Bodies, Representative of civic bodies, Chamber of Commerce & Industry and Policy Makers are the various target groups/stakeholders. These target groups can be classified into four categories in a developing economy. First category comprises nearly one tenth of the population who are proactive and ready to accept new ideas and generally impound upon every given public sector opportunity; second category comprises of next two fifth of the population who have doubts, linger on and try to catch up as tail enders or become self sufficient at their own pace; third category comprising of next two fifth of the population is the one reluctant to accept the programmes and want to catch up when the programme is over and the fourth category are the remaining one tenth of the population who are just antagonistic to any change bound by traditions and conservative approach and may ever be prone to remain backward. Different extension methodologies and strategies shall have to be evolved to address the aforementioned strata of the populace.

Brief review of the status of previous research extension activities

AFRI has carried out extension works through multifarious approaches encompassing organisation of educational / desert learning / eco-sensitisation / capacity building / technology transfer training programmes; Ph.D. programmes; consultancies; technical advice; evaluation studies; collaborative forestry research extension programmes; exhibits display of activities; publications of extension bulletins / handouts, brochures, leaflets, periodic newsletters, pamphlets and paper presentations / publications in print media; audio-visuals; T.V. & radio talks over electronic media; development of media action plan; organizing workshops / seminars / conferences / fairs; planting stock improvement, SPA, SSPA, CSO and VMG establishment; protocols of research pursuits and achievements; availability of planting propagules and on farm trials for the benefit of end users / stakeholders but not all at a time or in continuity. Systematic concerted effort has to be made on the status of comprehensive research extension on sustainable basis.

Notable had been training imparted to 3703 stakeholders under UNDP and DDP; training to 317 stakeholders in Gujarat under the NOVOD Board on neem; Capacity building and Eco-sensitisation of 2474 Stakeholders in ten desert districts of Rajasthan under CDP; Desert Learning Programme sponsored by Ashoka Trust for Research in Ecology and Environment; participation since 1998 in Swadeshi Melas and Western Rajasthan Hastshilp Utsav at Jodhpur; farmers' fare in 2000 under RDP; established demonstration trials at Jodhpur, Churu, Jasol (Barmer), Bikaner, Nagaur and Palanpur under the Ministry of Rural Development Project; Survey of 10 villages in two clusters of Jodhpur District for On Farm Trials under UNDP Project; site-specific Agroforestry trials on the farmer's land in two watersheds of Jodhpur Distt. under NABARD project; Media Action Plan for the Institute is in vogue; Radio talks through All India Radio Jodhpur; participation in the "chaupal" programme of the All India Radio to mitigate the problems faced by the farmers; Audio visuals / Electronic Media / films on Rain Water Harvesting and Khejri Mortality; development of Extension & Interpretation Centre at AFRI with depiction of a promising future; thirty ongoing projects; linkages and collaboration with of Water resources: implementation of project on Biodrainage; TERI: Formulation of GEF-UNDP Project; Department of Science and Technology: *Jatropha* mission; Department of Biotechnology: Multilocational trial of Bamboo and preparation of 'Operational Guidelines for Projects Supported by DBT', Collaborative projects with State Forest Departments of Rajasthan and Gujarat; IUFRO, APAFRI, CIFOR, IDRC, NMPB etc.; celebration of National Science Day, Van Mahotsava Day, World Environment Day etc.; publication of Brochures namely Arid Forest Research Institute Jodhpur, Know Your Nursery Pests and Their Control, Know Your Nursery Diseases and Their Management, Marwar Teak : Rohida (*Tecomella undulata*), Kalpavriksha of the Desert : Khejri (*Prosopis cineraria*), अरेबिक गोंद का स्रोत : कुमटा वृक्षए लवणीय भूमि का मित्र : जालए मरु भूमि का कल्प वृक्ष : खेजड़ी, मारवाड़

का सागवान : रोहिड़ा, बंजर भूमि की हरी चादर : विलायती बबूल, वर्षा ऋतु आई वृक्ष लगाने की रुत लाई, कैर : मरुस्थल में परती भूमि सुधार हेतु फल वृक्ष, शुष्क वन अनुसंधान संस्थान जोधपुर; publication of pamphlets on Medicinal Plants namely *Cassia angustifolia, Chlorophytum borivilianum, Emblica officinalis, Commiphora wightii, Withania somnifera, Tinospora cordifolia, Rauwolfia serpentina, Aloe vera, Plantago ovata, Phyllanthus amarus, Ocimum santum, Jatropha curcas, Glycyrrhiza glabra, Asparagus racemosus, Andrographis paniculata* and leaflets on रतनजोत, सर्पगंधा, कालमेघ, भूंई ऑवला, सफेद मूसली, गिलोय, ग्वारपाठा, शतावरी, गुग्गुल, खेजड़ी, तुलसी, कैर, अष्वगंधा, रोहिड़ा, कुमट, मुलैठी, इसबगोल, सोनामुखी।

Methodology and Strategies for developing forestry research extension technology

Involving developing threshold technologies lab to land programme characteristics:

- Review present extension strategies and thereby, devise parameters, criteria, indicators and benchmark of performance.
- Strive to garner support and funds for establishing 50 ha each Forestry Research Extension Centres in hyper arid, arid and semi-arid areas in lieu of none with the Institute.
- Associate and coordinate institutional integrated approach, identify and formulate viable long-term projects and diversification schemes for establishing and developing participatory demonstrative models, motivational diversification thereof.
- Devise forestry research pursuits to mitigate any adverse environmental and ecological effects.
- Explore suitability and adaptability, strengthen mobilization and specialization of extension expertise through training and refresher course packages, mass awareness programmes periodically in order to develop benchmark

services and standardized byproducts value addition entrepreneurship cooperatives.

- Strive for technical know-how acquaintance of stake-holders through mobile public awareness exhibits and publicity through multimedia.
- Organizing training and capacity building programmes through talks, educational tours, interactions and field visits on forestry research extension facilities, activities and services.
- Compile and issue documented guidelines on best practices of forestry research extension followed in various parts of the country.
- Provide for a qualitative resource base of forestry research standardization extension and
- Optimum inputs with maximizing output development through technological advancement, cost effective and fast track solutions.
- Legal bindings / agreements and insurance to ensure incentives for local groups and individuals amongst stakeholders to undertake afforestation or forest management. Assimilate standardized productive techniques of concluded experiments / trials / projects, mobilise and compile a resource base of data and information of plausible forestry research extension technologies in Library Information System Extension Centre as an ultramodern facility answerable to a common man for knowledge base of when, where, why, what and how about extension of institutional activities, developed forestry research extension technologies and resources. This system shall be pivoted upon institutional scientific background, administrative setup, natural and manmade resources. It shall place on the Internet scientifically and technically analyzed forestry research expertise and specialization information based inferences/ findings of the Institute. Forestry research extension reference resource base shall also be evolved in it from books,

newspapers, magazines etc. and updated burning topic wise all time solutions shall be available in the LIS, which shall be publicized for increasing knowledge of stakeholders. It will also disseminate available technical collaboration and assimilated institutional, national and global technological inputs for research extension in order to develop ecological and environmental resource base information centre of threshold technologies. It shall project the multifarious and multidimensional picture, as well as, information on assets and infrastructure facilities of the institute, which shall be conducive in formulation and implementation of integrated research, extension and development projects and schemes, and thereupon, serve as a source of multipurpose utility. Chronological figures of regional remote sensing centre shall also be integrated in this system. It shall provide forestry research extension facilities, activities and services encompassing time management, mass awareness, education and training at the doorstep of stakeholders through on-line E-governance, video-conferencing, remote sensing services and satellite based application centre. Anyone shall be able to access forestry research extension by the press of a button through an affordable price tag.

- Salient features of the Media Action Plan for Research Extension and Development shall include increased interaction with media personnel, a shelf of popular articles for periodic distribution, an electronic media visual bank on the success stories of AFRI, easy to understand interesting scientific terminology language for Media, local vernacular easy language research material, access to Press Information Bureau, streamlined press releases and dissemination through the central unit after approval of competent subject matter specialist and authority, periodic radio/television talks for wide coverage, mass media interaction with the local people for assessing socio-economic index of AFRI forestry research extension popularity.

Role of Medi-culture (Silva-Ayurveda Model) in Agroforestry

The National Forest Policy, 1988 calls for bringing at least 20% area in plains and 60% area in hills under forest cover which averages to 33% of total national geographic area. The target is to take present 23% cover to 27% by 2007 and 33% by 2012. But the shrinking small land holdings are beset with biotic pressure and burgeoning human population to involve effective intervention in areas outside forests. Hence, the solution to this situation lies in soil and moisture conservation along with vegetational cover through afforestation, carbon sequestration sustenance, low cost technology for vermi-compost, super-compost, VAM Fungi inoculation, increased commercial production of hybrid seeds, productive intercropping of medicinal, horticulture, silviculture and agricultural crops in tandem to cater to food, health, shelter, fodder and timber requirements of exploding population. Medi-culture of therapeutic utility, commercial viability, eco-friendly and cultivability hold scope for sustainable development through addressing the environment, employment and health concerns of the growing population. Appropriate plant part extraction with viability range maintenance, active ingredients/principles phyto-sanitary analysis, grading, storage, qualitative certification, knowing and drawing linkages for appropriate pricing and marketing mechanism, and value added pharmacopoeia entrepreneurship cooperatives is imperative. The cultivators have pure stands of a few traditional and few introduced silviculture/horticulture species. Trees/ bushes act as shelterbelts, micro-wind break, boundary demarcation, provide protection from livestock, conserve soil moisture and also provide conducive humid and shade conditions for symbiotic vegetative associations in dry arid region of northwestern India. Their increased cover shall also harbinger more rains. Integrating cultivation of most remunerative and productive vegetative measures, multi-purpose forestry/horticulture trees / bushes, medicinal herbs and climbers with site specific soil conservation and rainwater harvesting measures, and thereby, establishing their multi-tier demonstrative Silva-Ayurveda (Medi-culture) prototype models in arid region's problematic stress sites, especially, degraded pasture/ grazing lands, inland-sands, arid and semi-arid areas, saline–alkaline areas, waterlogged and marshy areas, advancing desert sites, undulating degraded hilly tracts along with their preservation for further improvement, phyto-products value addition and diversification for mass scale agroforestry to educate rural poor in order to generate additional income and raise their standard of living is the need of hour, further dependent upon the choice of species and number of species as per adaptable design.

Species recommended in order of priority

1. For advancing desert sites, semi-arid & arid areas

a. Trees

Top Storey: *Emblica officinalis* Kanchan Gaertn. (Aonla), Grafted *Ziziphus* spp., *Cordia myxa* / ruthii (Gunda/ Gundi), *Aegel marmelos* (Bil), *Moringa olifera / conconensis* (Sahjan), *Tecomella undulata* Seem. (Rohida), *Antocephalus cadamba* (Kadam), *Azadirachta indica* (Neem), *Prosopis cineraria* (Khejari), *Collaphospermum mopane*, *Hardiwickia binata* (Anjan), *Phoenix dactylifera* Halawy (Khajur).

Middle Storey: *Maytenus emerginata*, *Calligonum polygonoides* (Phog), *Acacia senegal* (Kumta/Khairi), *Acacia bevinosa*, *Balanites aegyptiaca*, *Dichrostachys cineraria*, *Punica granatum* Jalore seedless, *Melia azedarach* (Bakain).

b. Shrubs

Commiphora wightii (Arn.) Mukul/Bhandari (Guggal), *Sida cordifolia* (Mahabal), Grafted *Citrus* spp. (Neembu) *Capparis decidua* (Forssk.) Edgew (Ker), *Grewia tenax, Jatropha curcas, Ricinus communis, Moringa Pterygosperma, Lemonum acidum* and *Carissa coranda*.

c. Climbers/Lianas

Asparagus racemosus Willd (Satawari), *Gloriosa superba* Linn. (Kalihari), *Leptadenia reticulate* (Jiwanti), *Tinospora cordifolia* (Willd.) Hook/(DC) Miers (Guduchi/ Amrita), *Gymnema sylvestre* R.Br. (Gurmar), *Paederia foetida* (Prasarni) and *Tylophora asthematica* (Damabel), *Cissues quadrangularis, Mucuna pruriens, Piper longum, Momordica charautia*.

d. Herbs

Chlorophytum borivillanum Sant/ Baker (Safed Musli), *Anacyclus pyrethrum* DC. (Aquarqarha), *Withania somnifera* (Linn.) Dumal (Ashwagandha), *Convovulus pluricaulis* (Sankhpushpi), *Aloe vera* (Gwarpatha), *Curcuma longa, Cuminum cyminum*.

e. Grasses

Cenchrus ciliaris (Dhaman), *Lasiurus sindicus* (Sevan), *Aristida funniculata, Dactylotinium sindicum, Cympogon jwarncussa*.

2. For saline-alkaline areas

a. Trees

Acacia ampliceps, Acacia colei, Salvadora persica (Peelu), *Moringa oleifera* Lamk. (Sahjan), *Terminalia arjuna* (Arjun).

b. Shrubs

Atriplex lentiformis, Atriplex amnicola, Haloxylon salicornicum (Sazi), *Pluchea lancelata, Tamarix dioca, Capparis decidua* (Forssk.) Edgew (Ker).

c. Climber / Liana

Ipomea biloba and

d. Herbs

Cressa cretica (Rudrawanti), *Peganum hermala* (Hermal), *Acanthus ilicifolius* (nirmuli).

3. For water-logged areas

a. Trees

Eucalyptus camaldulensis, Acacia nilotica, Dalbergia sissoo.

b. Shrubs

Tamarix aphyla (Jhabuk), *Typha angustifolia, Rhizophora mucronata.*

c. Climber / Liana

Ipomea aquatica and

d. Herbs

Acorus calamus (Bachh), *Bacopa monnieri* (Linn) Pannell /Corr. (Brahmi).

4. Mulching

Crotolaria burhia, Aerva pseudotomentosa, Leptadenia pyrotechnica, Sesbanea aculeate/sesban (organic).

5. Live hedge

Acacia senegal, Leptadenia pyrotechnica, Clerode-ndron phlomoides, Acacia jacquemontii, Calligonum polygonoides, Opuntia spp., *Ipomea carnea, Lycium barbarum, Zizyphus nummularia, Prosopis juliflora, Euphorbia caducifolia, Euphorbia tirculli.*

6. Silva-pastoral

(a) Trees: *Zizyphus mauritiana, Ailanthus excelsa, Colophospermum mopane, Acacia bevinosa.*

(b) Grasses: *Cenchrus ciliaris, Lasirus sindicus, Aristida funniculata, Dactyloctinium sindicum.*

7. Sewage water utilization

Eucalyptus camaldulensis, Acacia nilotica and *Azadirachta indica.*

Design of Silva-Ayurveda Model

It is recommended to adopt a *Bush* line on mound bunds spaced 1 M. apart whose inter-spaces are to be filled by *thorny herb* planting at 30 cms. apart along a trench boundary specification 400 X 1.2 M. X 0.8 M. with mound made inside for protection of a multi-tier model with alternate tree and bush base plant species spaced 6 X 6 M. for arid region and 8 X 8 M. for semi-arid region with herbal plants inter-cropping at an average spacing of 45 X 45 Cm. The climbers can be introduced in the fourth year alongwith base plants. Legend X = Bush, ⊙ = Tree, bush-based: 6 X 6 M.; tree based: 8 X 8 M. and 6 X 6 M. respectively design is depicted below:

In the past, the Research Trials and Models had been at 6 X 6 M spacing. Hence the variations to suit the arid region in particular and others in general for long run have been recommended. It is also recommended that keeping in view the quantity of water availability year round, root-shoot competition and canopy competition, planting is to start either from a corner enclosure or from middle one tree species and plantation extension in any side or outward respectively as per optimum land use productivity enhancement can be undertaken in the aforementioned design. The different enclosures outward can have different or same tree/bush/climber/herb species.

MULTI-TIER SILVA-AYURVEDA MODEL

Linkages

It may be worthwhile to utilize the services of existing extension wings like Van Chetna Kendras, KVK's and NGO's with requisite subject matter specialization/ expertise/professionalism. The effort to transfer the technology ought to reflect a high benefit / cost ratio adaptable to a given environment. The technology may not reach to the desired target groups due to paucity of funds and other aforementioned communication gaps to make a lasting impact. The religious, cultural and socio-economic heterogeneity requires a flexible, open and eco-friendly blending approach for extension programmes. The effective technology transfer depends on an interactive and holistic information system, which includes a research sub-system; a dissemination sub-system and a user subsystem in tripartite agreement. After monitored dissemination of the technology, its utilization and evaluation is necessary so as to deliver the expected outcome. This process will not only sensitize planners and implementers to adopt productive progressive technologies in the field of forestry research extension but also lead to capacity building of all stakeholders. Research components along with monitoring and evaluation mechanism shall ever remain inbuilt in all the forestry development programmes including Forest Working Plans. A better management of our fragile ecosystem needs effective linkages for increasing biodiversity and ecosystem sustainability through:

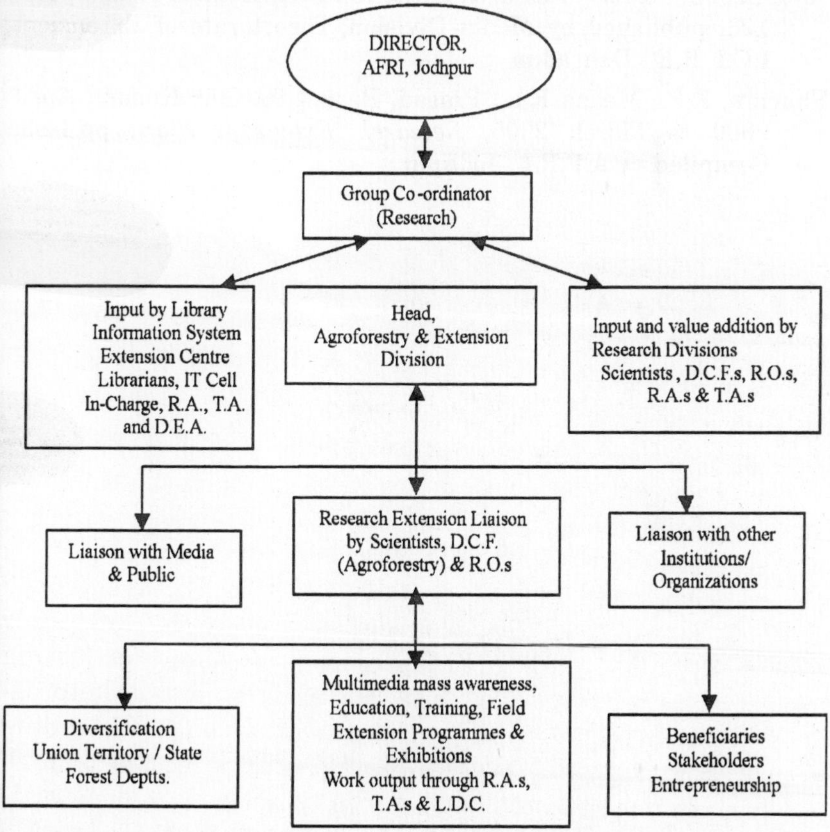

RESEARCH EXTENSION STRUCTURE OF ARID FOREST RESEARCH INSTITUTE, JODHPUR

Acknowledgements

The authors are thankful to the organisational and institutional research and technical personnel, facilities, activities and services vis-a vis stakeholders' expectations towards increasing scope of incentives for their more remunerative investments stakes as a source of inspiration towards writing this paper.

Literature Consulted

Anon, 2000, *Forestry Research Extension Programme ICFRE,* pp.1-106, published by Media Division, Directorate of Extension, I.C.F.R.E., Dehradun.

Sharma, R.P.; Meena R.L.; Prasad, R. and Parihar Kusum; April 2000 to March 2005, *Research Extension Plan*, pp.1-63, Compiled by A.F.R.I., Jodhpur.

RECOMMENDATIONS

- A minimum of 5-10% of the total budget of research should be set apart for Forestry Extension.
- Forestry Research Extension infrastructure development at district level.
- Linkages with KVK / Offices physically located within the KVK. Forestry research extension experts should be placed in KVK to reduce the expenditure.
- MOU between ICFRE and ICAR for extension interface.
- Forestry research extension should lead to assured market for stake holders with appropriate mechanism.
- Strengthening of extension wings of ICFRE institu-tions with placement of a sociologist and an economist.
- Budgetary provision for forestry research extension under R&D projects.
- Collaboration between stakeholders and scientists/ NGOs for identifying diagnosis, economics and desig-ning of forestry research extension requirements of user groups with bottom-up approach.
- Strengthening of capacity building and eco-sensitization of institutions.
- Some regulations for using treated wood of lesser known secondary plantation species.
- CPRs including Orans and Gauchars to be rehabilitated / strengthened / regenerated along with all the village institutions with dedication and proper technical inputs.
- Mine-over-burdens to be treated/re-vegetated/ rehabili-tated in a planned manner by appropriate agency in a scientific manner.

- Creation of research-cum-extension units at regional levels in the SFDs of Rajasthan, Gujarat and Dadar & Nagar Haveli with appropriate machinery.
- Strengthening of extension units with adequate man, material and machinery.
- Proper networking and sharing of knowledge.
- Proper monitoring and evaluation mechanism to be put in place.
- Proper protection and maintenance of research-cum-demonstration plots.
- Extension by kisan nurseries/farms.
- Effective implementation of suitable agro-forestry models.
- Demonstration trials of MFPs and Medicinal plants highlighting cultivation, processing, value addition and marketing.
- Remote sensing technology coupled with information technology and GIS to be utilized to augment extension efforts.
- Rejuvenation of water bodies with the help of stakeholders and proper technological interventions to increase water productivity.
- Provision for permanent demonstration plots in the mandated areas.
- Provision of normative expenditure for demonstration of technological packages.
- Establishment and development of site specific demonstrative models.
- Provision of Publicity Van and mobile exhibits.

संस्तुतियाँ

- अनुसंधान हेतु उपलब्ध कुल बजट का कम से कम 5 से 10 प्रतिशत भाग वानिकी विस्तार हेतु रखा जाना चाहिये।
- जनपद स्तर पर बुनियादी एवं मूलभूत सुविधाएँ विकसित की जानी आवश्यक हैं।
- कृषि विज्ञान केन्द्रों के साथ संपर्क स्थापित करके भौतिक रूप से वहाँ पर कार्यालय स्थापित करना तथा अनावश्यक व्यय को रोकने हेतु वहाँ वानिकी अनुसंधान एवं विस्तार विशेषज्ञ को पदस्थ करना।
- भारतीय वानिकी अनुसंधान एवं शिक्षा परिषद तथा भारतीय कृषि अनुसंधान परिषद के मध्य आपसी समझ स्थापित करना।
- वन विस्तार के माध्यम से उचित तंत्र के साथ (स्टेकहोल्डर्स) पर्णधारियों हेतु विपणन सुनिश्चित किया जाना चाहिये। एक समाजशास्त्री तथा एक अर्थशास्त्री की नियुक्ति करके भारतीय वानिकी अनुसंधान एवं शिक्षा परिषद के अंतर्गत कार्यरत संस्थानों के विस्तार एककों को मजबूत करना।
- अनुसंधान एवं विकास परियोजनाओं के विस्तार हेतु बजट प्रावधान सुनिश्चित करना।
- आधारभूत जानकारी सहित आर्थिकी, निदान एवं परिकल्पना की पहचान हेतु (स्टेकहोल्डर्स) पर्णधारियों, वैज्ञानिकों एवं गैर–सरकारी संगठनों के मध्य सहयोग स्थापित करना।
- क्षमता निर्माण करने वाली संस्थाओं को मजबूत करना।
- कम जानकारी वाली लघु रोपण प्रजातियों की उपचारित काष्ठ के उपयोग हेतु कुछ नियम तथा विनियम तैयार करना।
- समर्पण की भावना के साथ उचित तकनीकी क्षमता को विकसित करके सभी ग्रामीण संस्थाओं के ओरण एवं गोचर सहित सामुदायिक लोक संसाधनों का पुनर्वास करने के उपरांत उन्हें पुनर्जीवित करके मजबूती प्रदान करना।

- खदान कार्यों के कारण जो भूमि नष्ट हो गई है उस भूमि को उचित अभिकरणों (एजेन्सी) द्वारा योजनाबद्ध एवं वैज्ञानिक ढंग से उपचारित कर वहाँ अनुकूलतम वनस्पतियों एवं पौधों का रोपण करके उसका पुनर्वास सुनिश्चित करना।
- उचित कार्यतंत्र के राजस्थान, गुजरात, दादर तथा नगर हवेली के राज्य वन विभागों में क्षेत्रीय स्तर पर अनुसंधान एवं विस्तार एककों का निर्माण किया जाना।
- पर्याप्त जनशक्ति, साज–सामान एवं कार्यतंत्र के साथ विस्तार एककों को मजबूती प्रदान करना।
- उचित नेटवर्क तथा ज्ञान–विज्ञान का आदान–प्रदान करना।
- स्थान विशेष में उचित अनुश्रवण एवं मूल्यांकन तंत्र स्थापित करना।
- अनुसंधान एवं प्रदर्शनार्थ रखे जाने वाले भूखंडों की उचित सुरक्षा व्यवस्था एवं रखरखाव सुनिश्चित करना।
- किसान रोपणियों द्वारा विस्तार किया जाना।
- उचित कृषि–वानिकी प्रारूपों का प्रभावी अनुपालन सुनिश्चित करना।
- लघु वन उत्पादों एवं औषधीय पौधों की खेती, उनकी विभिन्न प्रक्रियाएँ, मूल्य संकलन एवं विपणन पर बल देते हुये उनका प्रदर्शन परीक्षण करना।
- सूचना प्रौद्योगिकी तथा वैश्विक अंतर्राष्ट्रीय तंत्र के साथ सुदूर सुग्राही प्रौद्योगिकी को जोड़कर उनके बेहतर उपयोग हेतु प्रभावी प्रयत्न करना।
- जल उत्पादकता में बढ़ोतरी लाने के उद्देश्य से उचित प्रौद्योगिकीय हस्तक्षेप द्वारा (स्टेकहोल्डर्स) पर्णधारियों की सहायता से जल निकायों को मजबूती प्रदान करना।
- निर्धारित क्षेत्रों में स्थाई प्रदर्शनकारी भूखंडों का प्रावधान सुनिश्चित करना।
- प्रभावकारी प्रौद्योगिकीयों हेतु मानकीय आधार पर व्यय किये जाने का प्रावधान सुनिश्चित करना।
- स्थल अभिमुखी प्रदर्शनीय प्रारूप तैयार करना।
- प्रचार वाहिनी एवं चलती–फिरती प्रदर्शनियों का आयोजन करना।